DEVELOPMENTALISM IN EARLY CHILDHOOD AND MIDDLE GRADES EDUCATION

CRITICAL CULTURAL STUDIES OF CHILDHOOD

Series Editors:
Marianne N. Bloch, Gaile Sloan Cannella, and Beth Blue Swadener

This series will focus on reframings of theory, research, policy, and pedagogies in childhood. A critical cultural study of childhood is one that offers a "prism" of possibilities for writing about power and its relationship to the cultural constructions of childhood, family, and education in broad societal, local, and global contexts. Books in the series will open up new spaces for dialogue and reconceptualization based on critical theoretical and methodological framings, including critical pedagogy, advocacy and social justice perspectives, cultural, historical and comparative studies of childhood, post-structural, postcolonial, and/or feminist studies of childhood, family, and education. The intent of the series is to examine the relations between power, language, and what is taken as normal/abnormal, good and natural, to understand the construction of the "other," difference and inclusions/exclusions that are embedded in current notions of childhood, family, educational reforms, policies, and the practices of schooling. *Critical Cultural Studies of Childhood* will open up dialogue about new possibilities for action and research.

Single authored as well as edited volumes focusing on critical studies of childhood from a variety of disciplinary and theoretical perspectives are included in the series. A particular focus is in a re-imagining as well as critical reflection on policy and practice in early childhood, primary, and elementary education. It is the series intent to open up new spaces for reconceptualizing theories and traditions of research, policies, cultural reasonings and practices at all of these levels, in the USA, as well as comparatively.

The Child in the World/The World in the Child: Education and the Configuration of a Universal, Modern, and Globalized Childhood
   Edited by Marianne N. Bloch, Devorah Kennedy, Theodora Lightfoot, and
   Dar Weyenberg; Foreword by Thomas S. Popkewitz

Beyond Pedagogies of Exclusion in Diverse Childhood Contexts: Transnational Challenges
   Edited by Soula Mitakidou, Evangelia Tressou, Beth Blue Swadener, and
   Carl A. Grant

"Race" and Early Childhood Education: An International Approach to Identity, Politics, and Pedagogy
   Edited by Glenda Mac Naughton and Karina Davis

Governing Childhood into the 21st Century: Biopolitical Technologies of Childhood Management and Education
   By Majia Holmer Nadesan

Developmentalism in Early Childhood and Middle Grades Education: Critical Conversations on Readiness and Responsiveness
   Edited by Kyunghwa Lee and Mark D. Vagle

DEVELOPMENTALISM IN EARLY CHILDHOOD AND MIDDLE GRADES EDUCATION

Critical Conversations on Readiness and Responsiveness

Edited by

Kyunghwa Lee and Mark D. Vagle

2010

ISBN 978·0·230·61980·7 palgrave
macmillan

First published in 2010 by
PALGRAVE MACMILLAN®
in the United States—a division of St. Martin's Press LLC,
175 Fifth Avenue, New York, NY 10010.

Where this book is distributed in the UK, Europe and the rest of the world,
this is by Palgrave Macmillan, a division of Macmillan Publishers Limited,
registered in England, company number 785998, of Houndmills,
Basingstoke, Hampshire RG21 6XS.

Palgrave Macmillan is the global academic imprint of the above companies
and has companies and representatives throughout the world.

Palgrave® and Macmillan® are registered trademarks in the United States,
the United Kingdom, Europe and other countries.

ISBN: 978–0–230–61980–7

Library of Congress Cataloging-in-Publication Data is available from the
Library of Congress.

A catalogue record of the book is available from the British Library.

Design by Newgen Imaging Systems (P) Ltd., Chennai, India.

First edition: June 2010

10 9 8 7 6 5 4 3 2 1

Printed in the United States of America.

CONTENTS

Series Editor's Preface

Over the past fifty years, especially as related to civil rights gains in the 1960s, discourses of diversity, equity, contingency, and possibility have emerged from a range of cultural, intellectual, and grassroots locations. These discourses and ways of understanding apply to all of us, those who are younger as well as those who are older, those who have been labeled children as well as those labeled adults. A critical social science has emerged across fields such as education, sociology, and even medicine as new transdisciplinary fields such as cultural and gender studies have appeared. Postmodern perspectives have challenged universalist grand narratives, whether related to life conditions such as poverty or constructions of learning and human change. Feminist perspectives have revealed the importance of a multiplicity of life journeys, while at the same time unmasking the gendered power that has, in many cases, dominated those journeys. Critical and indigenous voices of understanding have made known the interstices of privilege and oppression embedded within life experiences. These discourses have made public the infinite ways in which life is experienced and interpreted. Although these diverse voices have gained attention, the possibilities that they generate have often been denied through misunderstanding, fear of loss of power, and the reemergence of dominant orientations. The continued use of universalist developmental constructs regarding those who are younger is one of those dominant orientations.

The editors and authors of this volume join the diverse critical voices who would "poke some holes in the lid" of developmentalism to provide air because the "new and critical perspectives are suffocating." The volume addresses the commonalities in the fields of early childhood and middle grades education that have resulted from the reliance on the grand narrative of developmental theories. In the first section "Are They Ready Yet?" the authors focus on the construct of "readiness" as defined in multiple locations, and the ways that the construct separates sense-making from life experiences and denies complex abilities. Notions such as an ethical answerability in the moment and mindfulness are introduced as alternatives to constructions of readiness that embody deficiency and are always future oriented. In the second section "Responsive to What?" the authors demonstrate the binaries that are reinforced through constructs such as "responsiveness." Rather than perpetuating a developmentally embedded (and authoritarian) practice such as response, they discuss the construction of contingent and recursive relations that involve all in the struggle to learn together. Finally, the last two chapters represent conversations across fields (of early childhood and middle grades education) and suggest that developmentalists and critical social scientists (e.g., reconceptualists, feminists, critical multiculturalists) share commitments to serving the needs of those who are younger and that this sharing could lead to a reimagining of the fields.

Gaile S. Cannella, Series Editor
Professor and Velma E. Schmidt Endowed
Chair of Early Childhood Studies
University of North Texas

Developmentalism and the Need for Critical Conversations within and across the Fields

Kyunghwa Lee and Mark D. Vagle

Overview

As scholars in the fields of early childhood and middle grades education, we have been struck by common issues across our fields—most of which have resulted from the fields' reliance on grand developmental theories prevalent throughout the past century. By assuming universal stages and age norms, developmental perspectives have portrayed humans as perpetually being on their way toward something and as simultaneously lacking in reference to the intended goals. One consistent thread that runs through schooling practices in these fields has been an assumed developmentalism, leveraged through discourses of developmental appropriateness and readiness in early childhood education and developmental responsiveness in middle grades education.

To date, developmentalism—a heavy, dominant discourse—has served as the unwavering justification for the two fields. In fact, it can be argued that the fields would not even exist if early childhood and young adolescence were not treated as clearly defined stages of human development. In this respect, it is a bit dangerous to critique the very foundation on which early childhood and middle grades education stands. This could, however, be the very same reason to propose such critiques. It seems important for fields to continually bring multiple perspectives to bear on foundational matters. To this end, the purpose of this book is to facilitate a dialogue about the ways in which these universal and linear notions of development reinforce each other across fields and how the fields can be continually reimagined using multiple perspectives on growth and change. We situate this book in larger efforts to explore various frameworks that might help us move beyond the constraint of the past century's grand developmental theories and broaden the fields' vision for educating young people at particular moments in their lives.

In what follows, we discuss how developmentalism has been contextualized in each field through the notions of developmental appropriateness/readiness (early childhood) and developmental responsiveness (middle grades).

EARLY CHILDHOOD CONTEXT

The idea that education should follow the natural order and laws of child development has a long history in the field of early childhood education. Textbooks discussing early childhood history trace back to Plato as the first person who emphasized the importance of education during the early years of human life and describe Comenius, Rousseau, Pestalozzi, and Froebel as forerunners who recognized young children's distinct characteristics and proposed the idea that educational experiences for these children should be organized differently than those provided for older learners or adults (Braun & Edwards, 1972; Nourot, 2005). Stages of human development appeared in Rousseau's and Froebel's works (Wolfe, 2000). In particular, Froebel, the founder of the kindergarten in 1840, conceptualized childhood as having two periods, "early childhood (from birth to age 8) and later childhood or the 'scholar period'" (Walsh et al., 2001, p. 98), and strongly influenced the field's identity as a profession for children from birth to age eight.

The notion of education based on children's development has drawn educators' attention particularly since the child study movement led by G. Stanley Hall at the end of the nineteenth century. Weber (1969), describing kindergarten teacher education at the turn of the century, wrote, "It was argued that teachers working with young children needed an extensive understanding of human development. Indeed, Hall maintained the need for psychological knowledge 'increased inversely with the age of the student'" (p. 120). Hall's child study influenced many educators, including Patty Smith Hill who accepted developmentalism as "scientific" (Chung & Walsh, 2000) and invited leading educators and psychologists to the National Committee on Nursery Schools (NCNS) organized in 1926. Among the committee members was Arnold Gesell, a student of Hall's, who established and promoted norms for child development. The NCNS became the National Association for Nursery Education in 1929 and was renamed in 1964 as the National Association for the Education of Young Children (NAEYC), the field's leading professional organization to this day (NAEYC, n.d.; Wolfe, 2000). Child development knowledge has been a significant source for teacher education throughout the twentieth century to make the field "scientific and professional" (Bloch, 1992, p. 9). Freud, Erikson, and Piaget are among the most influential psychologists who have affected the field's perceptions of children as well as the design of educational curricula and programs for young children (Cannella, 1997).

Reflecting the field's reliance on developmental theories for its vision for early schooling, NAEYC published the first edition of the position statement, *Developmentally appropriate practice in early childhood programs serving children from birth through age 8*, in 1987 (Bredekamp, 1987). This position statement known as DAP among early childhood educators stated, "Human development research indicates that there are universal, predictable sequences of growth and change that occur in children during the first 9 years of life" (p. 2) and emphasized that educational experiences for young children should be age- and individual-appropriate. The guidelines then described what could be considered appropriate practices and what could be inappropriate. The position statement clearly revealed its endorsement of the past century's universal and linear stage theories of human development. Although the subsequent editions of the position statement published by NAEYC in 1997 and 2009 dropped the term "universal," and added the importance of considering the social and cultural contexts, the revised guidelines still maintained the notion of development as "relatively stable, predictable sequences of growth and change" (Bredekamp & Copple, 1997, p. 10; Copple & Bredekamp, 2009, p. 11) and emphasized the teacher's "knowledge of how children within a given age span typically develop and learn" (ibid.) for the design of the educational environment and curriculum.

The discourse of developmental appropriateness reveals the field's core beliefs about developmental stages and biological maturation. Many early childhood educators "see development as occurring in stages. When they speak of children and learning, they speak of 'readiness'" (Walsh, 1991, p. 113). In this context, developmental appropriateness and readiness constitute each other as both rely on stage theories and age norms. Graue (1992) also argued that conceptions of readiness "share a legacy of psychological thought that has permeated the child development/early childhood education field in the twentieth century" (p. 63). As powerful constructs, developmental appropriateness and readiness influence many areas in the lives of educators and children in early schooling.

The publication of the NAEYC's first position statement has prompted heated discussions and debates among early childhood educators and researchers about meanings of child development and politics involved in determining appropriate practices (e.g., Kessler, 1991; Lubeck, 1994; Walsh, 1991). In particular, a group of scholars identified as reconceptualists have critiqued the dominance of developmentalism and challenged the field to explore alternative perspectives (Bloch, 1991; Cannella, 1997; Kessler & Swadener, 1992; Mallory & New, 1994; Ryan & Grieshaber, 2005). Although these efforts have brought healthy skepticism to the field's assumed consensus and unquestioned orthodoxy, we feel as if the field has been divided into two major camps over the years: one for developmentalists and the other for reconceptualists as reflected in the presence of two Special Interest Groups (SIGs) of the American Educational Research Association, the Early Education and Child Development SIG and the Critical Perspectives on Early Childhood Education SIG.

We believe that a critical examination of developmental appropriateness and readiness and the exploration of various frameworks among scholars across disciplines and fields presented in this book will contribute to the early childhood field's ongoing efforts to move beyond the limitation of the dominant grand developmental theories by bringing additional issues and ideas to consider. We also hope that this book can promote dialogues, rather than divisions, among scholars with differing orientations and perspectives to continuously reimagine the field.

MIDDLE GRADES CONTEXT

Middle grades educators talk less about developmental appropriateness and readiness and more about developmental *responsiveness*—assuming that the developmental needs of young adolescents are particularly unique and therefore must be responded to in equally particular ways (National Middle School Association, 1982, 1995, 2003).

As Vagle discusses in chapter six, the creation of middle grades education indeed depends on a developmental depiction of adolescence, which dates back most notably to the writings of G. Stanley Hall (1904) and has been solidified throughout the last century in psychology through the work of Piaget (1952, 1960), Flavell (1963), and Erikson (1963). Conversations about adolescence both deepened and widened over this time. Throughout the 1900s those interested in the schooling of those in the "younger years" of adolescence advocated for something unique. John Lounsbury and Gordon Vars (2003) trace this progression back to 1909 when the first junior high school, Indianola Junior High School in Columbus, Ohio, was created. Lounsbury and Vars continue their historical trace by noting that nine years later, the Commission on the Reorganization of Secondary Education in its report *Cardinal Principles of Secondary Education* recommended the six-three-three organization of schools (i.e., one–six; seven–nine; ten–twelve)—in effect establishing a unique school for young adolescents. By 1945 the six-three-three pattern became the majority practice. In 1963, William Alexander (reprinted in 1995) advocated the term *middle school* in a speech at Cornell University. A decade later National Middle School Association (NMSA) was established and has, according to Lounsbury and Vars, since served as the leading U.S. organization dedicated to the unique needs of young adolescents. The NMSA now has affiliate organizations in Canada, New Zealand, Australia, and throughout Europe. Over the past thirty-five years, the five-three-four pattern (i.e., one–five; six–eight; nine–twelve) has become the most common school organization structure in the United States.

That said, many middle grades advocates, which Mark considers himself, are quick to point out that middle schooling is less about school structures and more about the unique ways in which schooling for young adolescents should be enacted. *This We Believe*, a seminal position paper published by NMSA, sets forth a clear vision for the schooling of young adolescents. At the heart of this vision is the young adolescent and developmental responsiveness.

A linear, unidirectional, and time-bound conception of development has dominated discussions in the field (Brown & Saltman, 2005; Lesko, 2001). We feel that a serious consideration of critical perspectives regarding young adolescence will help push back against stage developmentalism and in turn breathe some much needed air into middle schooling. One of our hopes is that this book will serve as one possbile way for other perspectives to be considered. It follows in the footsteps of Enora Brown and Kenneth Saltman (2005) as they incited, in our estimation, a hopeful dialogue about the challenges and possibilities of trying to use critical theoretical perspectives to imagine what (generous, loving, complicated, conflicting, contextual) schooling for young adolescents might be like. They open their edited book *The Critical Middle School Reader* as follows:

> Our intent is to create a space for dialogue and thinking about the origins of, goals of, and assumptions about the middle school concept, about youth, and about the future direction for an educational process that can and should promote personal, social, and societal transformation. This book is a starting point, since it is the *first* critical middle level reader, and we hope that it will inspire a broader conversation about the future of middle school and the possibilities for it to be reconceptualized through hopeful social ideals. (P. 8; emphasis in the original)

We feel that Brown and Saltman's (2005) reader, and in turn this book, could not have come at a better time. By continuing to rely solely on a linear, unidirectional, and time-bound conception of development and without serious consideration of alternative critical perspectives regarding adolescence, middle grades education runs the risk of stagnating and failing to achieve its ultimate goal—to create the best schools possible for ten- to fifteen-year-olds.

BOOK STRUCTURE

The book is organized in three sections. The first two sections contain conceptual essays that explore big ideas in relation to developmentalism and examples of empirical work that examine some particular challenges of using theories that move beyond the dominant grand developmental perspectives with regard to notions of readiness and responsiveness in early childhood and middle grades education. The third section includes cross-field conversational essays that discuss both common and unique issues related to developmentalism in the fields of early childhood education and middle grades education. Together, these chapters illustrate a variety of perspectives, paradigms, and approaches to the dominant developmentalism, including Bakhtinian, Buddhist, cultural psychologist, and post-structuralist frameworks.

Although multiple perspectives are represented, this work is by no means fully inclusive. For instance, we do not have writings that situate discussions of early childhood and middle grades schooling in market-based, capitalist contexts. There is much to be gained by in-depth analyses of how children are educated as commodities that can be bought and sold in the market or at the very least as participants contributing to and subjected to global capitalism. We also realize that the particular place in which we find ourselves historically—in the worst recession since the Great Depression—makes such contexts and the associated challenges even more acute. In this same manner, we also realize that we do not have a historian's perspective—one that can help all of us make sense of what was happening at given periods of time and to make some sense of where we are now.

However, despite these and certainly other omissions, we feel that this text is significant as schools in the United States have become increasingly high stakes testing environments since the 1980s when the test-based educational reforms combined with the economic justification linking poor student achievement to the decreasing U.S. competitiveness in the global markets began greatly influencing schooling practices (Shepard, 1999). As exemplified by the *No Child Left Behind (NCLB) Act* of 2001, it is difficult to keep young people at the forefront of policy and practice when their test score becomes the primary way they are talked about and schooled. Although we see the critical alternatives to grand developmental theories that we present as opportunities to push back against stage developmentalism, we have also tried to not be dismissive of developmental perspectives.

We heed James Beane's (2005) warning that as long as developmentalists and critical theorists "remain out of touch with each other, both are diminished" (p. xv). This was difficult to accomplish, however, as either-or logic is pervasive in such arguments. That is, in debates about developmental and critical perspectives, neither side seems to want to acknowledge the other, as presumably this would mean "losing" or reinforcing the very stance one is trying to disrupt. Although most of the individual chapters in this book tend to "take a side," collectively we have aimed to include multiple perspectives that can be considered alongside one another. We were especially conscious of this commitment in our coauthored chapters in section three.

An image we want to forward is one that Vagle has theorized in his research and teaching. Working from Lesko's (2001) attempts to "chip away at the Teflon coating" (p. 192) that is the grand developmental theories, Vagle pictures a box with a lid placed on top of it. Inside the box is where innumerable new and critical perspectives reside. The lid is the dominant developmentalism. It is very thick— designed to keep all of its contents safe and controlled. Although the lid has been softened, it still does not have many (if any) holes to allow for air, like a hamster cage might. The new and critical perspectives are suffocating. Like Vagle, we also hope to poke some holes in the lid, thus allowing other perspectives to get some air, with the long-term goal of prying the lid away from its conflated relationship with childhood and adolescence—making it one of the innumerable possible perspectives instead of the dominant one.

References

Beane, J. A. (2005). Foreword. In E. R. Brown, & K. J. Saltman (eds), *The critical middle school reader* (pp. XI–XV). New York: Routledge.

Bloch, M. (1991). Critical science and the history of child development's influence on early education research. *Early Education and Development*, 2, 95–108.

———. (1992). Critical perspectives on the historical relationship between child development and early childhood education research. In S. A. Kessler & B. B. Swadener (eds), *Reconceptualizing the early childhood curriculum: Beginning the dialogue* (pp. 3–20). New York: Teachers College Press.

Braun, S. J., & Edwards, E. P. (1972). *History and theory of early childhood education*. Belmont, CA: Wadsworth.

Bredekamp, S. (1987). *Developmentally appropriate practice in early childhood programs serving children from birth through age 8*. Washington, DC: National Association for the Education of Young Children.

Bredekamp, S., & Copple, S. (Eds). (1997). *Developmentally appropriate practice in early childhood programs* (revised ed.). Washington, DC: National Association for the Education of Young Children.

Brown, E. R., & Saltman, K. J. (Eds). (2005). *The critical middle school reader*. New York: Routledge.

Cannella, G. S. (1997). *Deconstructing early childhood education: Social justice and revolution*. New York: Peter Lang.

Chung, S., & Walsh, D. J. (2000). Unpacking child-centeredness: A history of meanings. *Journal of Curriculum Studies*, 32, 215–234.

Copple, S., & Bredekamp, S. (2009). *Developmentally appropriate practice in early childhood programs serving children from birth through age 8* (3rd ed.). Washington, DC: National Association for the Education of Young Children.

Erikson, E. (1963). *Childhood and society*. New York: W.W. Norton.

Flavell, J. H. (1963). *The developmental psychology of Jean Piaget*. Princeton, NJ: Van Nostrand Reinhold.

Graue, M. E. (1992). Meanings of readiness and the kindergarten experience. In S. A. Kessler & B. B. Swadener (eds), *Reconceptualizing the early childhood curriculum: Beginning the dialogue* (pp. 62–92). New York: Teachers College Press.

Hall, G. S. (1904). *Adolescence: Its psychology and its relation to physiology, anthropology, sociology, sex, crime, religion, and education*. New York: Appleton & Company.

Kessler, S. A. (1991). Early childhood education as development: Critique of the metaphor. *Early Education and Development*, 2, 137–152.

Kessler, S. A., & Swadener, B. B. (Eds). (1992). *Reconceptualizing the early childhood curriculum: Beginning the dialogue*. New York: Teachers College Press.

Lesko, N. (2001). *Act your age: A cultural construction of adolescence*. New York: Routledge/Falmer.

Lounsbury, J. H., & Vars, G. F. (2003). The future of middle level education: Optimistic and pessimistic views. *Middle School Journal, 35*(2), 6–14.

Lubeck, S. (1994). The politics of developmentally appropriate practice: Exploring issues of culture, class, and curriculum. In B. L. Mallory & R. S. New (eds), *Diversity & developmentally appropriate practices: Challenges for early childhood education* (pp. 17–43). New York: Teachers College.

Mallory, B. L., & New, R. S. (Eds). (1994). *Diversity & developmentally appropriate practices: Challenges for early childhood education.* New York: Teachers College Press.

National Association for the Education of Young Children. (n.d.). *Historical overview.* Retrieved July 29, 2009, from http://www.naeyc.org/about/history.

National Middle School Association. (1982). *This we believe.* Westerville, OH: National Middle School Association.

———. (1995). *This we believe: Developmentally responsive middle level schools.* Westerville, OH: National Middle School Association.

———. (2003). *This we believe: Successful schools for young adolescents.* Westerville, OH: National Middle School Association.

Nourot, P. M. (2005). Historical perspectives on early childhood education. In J. L. Roopnarine & J. E. Johnson (eds), *Approaches to early childhood education* (4th ed., pp. 3–43). Upper Saddle River, NJ: Prentice Hall.

Piaget, J. (1952). *The origins of intelligence in children.* New York: International University Press.

———. (1960). *The child's conception of the world.* Atlantic Highlands, NJ: Humanities Press.

Ryan, S. K., & Grieshaber, S. (2005). Shifting from developmental to postmodern practices in early childhood teacher education. *Journal of Teacher Education, 56,* 34–45.

Shepard, L. A. (1999). The influence of standardized tests on the early childhood curriculum, teachers, and children. In B. Spodek & O. N. Saracho (eds), *Issues in early childhood curriculum* (pp. 166–189). Troy, NY: Educator's International Press, Inc.

Walsh, D. J. (1991). Extending the discourse on developmental appropriateness: A developmental perspective. *Early Education and Development, 2,* 109–119.

Walsh, D. J., Chung, S., & Tufekei, A. (2001). Friedrich Wilhelm Froebel. In J. A. Palmer, L. Bresler, & D. E. Cooper (eds), *Fifty major thinkers on education: From Confucious to Dewey* (pp. 94–99). New York: Routledge.

Weber, E. (1969). *The kindergarten: Its encounter with educational thought in America.* New York: Teachers College Press.

Wolfe, J. (2000). *Learning from the past: Historical voices in early childhood education.* Mayerthorpe, Alberta: Piney Branch Press.

PART I

Are They Ready Yet?

Kyunghwa Lee

As Graue (1992, 1993) illustrated in her study about meanings of readiness in the kindergarten context, the notion of readiness is "a dearly held construct" (1992, p. 93) in the field of early childhood education. Parents and teachers of young children ponder over whether their children are ready for entry into kindergarten or for the promotion to the next grade level. Readiness is also a topic of great interest to the general public and policymakers as reflected in its frequent appearance in the popular press and the political rhetoric. A series of quick Google searches conducted while preparing for this introduction revealed 18,800,000 results for the term "readiness"; 1,230,000 results for the combined terms "readiness and early childhood"; and 981,000 results for "readiness and young children." Sources of information posted on websites in relation to readiness vary, ranging from academic publications, government reports (e.g., documents posted by the National Governors Association), childcare organizations, the media (e.g., PBS teachers), to parent magazines, and so on.

In this section, scholars with expertise in early childhood education, mathematics education, and social studies education—all actively involved in early childhood or middle grades teacher education—critically examine the notion of readiness that they have encountered in their work. The purpose of this cross-discipline and cross-field examination is to identify common issues related to the construct of readiness across contexts and to explore alternative perspectives useful for research and practice within the field of early childhood education and across these related fields of teacher education.

The first two chapters in this section discuss how early childhood teachers deliberately or unwittingly rely on the construct and rhetoric of readiness for their decision-making about instruction and children. In the first chapter, *What's More Important: Numbers or Shoes? Readiness, Curriculum and Nonsense in a Rural Preschool,* Parks and Bridges-Rhoads examine, based on a six-month ethnographic study, how the discourse of readiness frames nearly all of the interactions between teachers and children in the observed preschool classroom. Parks and Bridges-Rhoads problematize how "the forward-looking" written curriculum (*SRA Language for Learning*) encouraged the teachers and children to engage in a number of practices that discouraged children's sense-making by separating real life experience from classroom learning and by disrupting meaningful conversations in genuine contexts. Those practices, however, were justified as preparing the preschoolers, who came primarily from low-income African American families, for kindergarten. Parks and Bridges-Rhoads discuss how critiquing this curriculum by using the notion of developmental appropriateness can still be limiting. Instead, they draw on Bakhtin's (1990, 1993) concept of "answerability" to explore the possibilities of a pedagogy that demands that educators be ethically accountable to children in the moment with them as opposed to a pedagogy centered on preparation for the future.

In the second chapter, *Who is Normal? Who is Abnormal? Rethinking Child Development from a Cultural Psychological Perspective,* I draw on data from my research on early childhood teachers' perceptions about children with attention deficit/hyperactivity disorder (ADHD) to illustrate how the notions of age-based developmental norms and readiness are used in the discourse and practice of identifying normal and abnormal child development. I critique the field's allegiance to an outdated and limited developmental perspective focusing on biology and the individual psychology separated from culture, a perspective no longer viable even in the field of developmental psychology. Looking at ADHD diagnosis and pharmaceutical treatment as a cultural practice, I propose cultural psychology (e.g., Bruner, 1996; Cole, 1996; Kitayama & Cohen, 2007; Shweder et al., 1998) as a contemporary developmental systems model (Lerner, 1998, 2006) that can help early childhood educators move beyond the limitations of the past century's grand developmental theories and consider the sociohistorical context influencing their perceptions about children's development, including disability.

Although prevalent, readiness is not confined to early schooling. The next two chapters in this section reveal how the construct and rhetoric of readiness is also influential in the contexts of middle grades and teacher education. In *Being Present in the Middle School Years*, Conklin helps us see how the future-oriented education permeates American schooling. Drawing on a semester-long case study, Conklin found that many secondary preservice teachers and their cooperating teachers perceived middle grades education and the period of young adolescence itself as "a stepping stone"—a stage to prepare for better learning and specifically for high school. Based on this data and other researchers' work, Conklin argues that the future-oriented education often prevents educators from providing students with intellectually challenging and engaging learning. She introduces Buddhist monk Thich Nhat Hanh's (1987a, b, 1993) notions of "living in the present moment" and "mindfulness" to explore how these ideas might encourage educators to reconsider their future-oriented curriculum and pedagogy and to look at the learners in the here and now. Conklin cautions that Hanh's ideas should not be interpreted "as a short-sighted method of 'living in the moment' without regard to future consequences." Instead, she argues that Hanh's ideas about paying careful attention to learners in the present moment help educators recognize young people as intellectually curious beings who are eager to engage in challenging learning.

In *Am I a Novice Teacher? The Voices of Induction Teachers in a Preschool*, Park and Parks situate the discourse of readiness in the context of teacher education and challenge stage theories of teacher development that position novices as peripheral participants in school communities and that portray induction teachers as being unready to handle more complex and challenging issues arising in teaching. Drawing on a qualitative case study with induction teachers working in a preschool in South Korea, Park and Parks illustrate the strengths novice teachers bring to thinking about and executing teaching practice. They argue that induction teachers play a role as social negotiators and resistors of imposed criteria and labeling such as "novice" or "brand new" in order to find and represent themselves in their everyday lives. By demonstrating the ways in which novice teachers resisted dominant thinking about children and schooling practice, Park and Parks ask us to reconsider the prevalent notion that novice teachers move in a linear way toward becoming expert teachers. Using a Bakhtinian (1993) perspective, these authors argue that novice teachers can be seen as acting from their unique places in Being and as qualitatively different from experienced teachers with particular perspectives and experiences, but not as necessarily moving toward a "better" stage in teaching.

In the last chapter of this section, *Responsivity rather than Readiness,* Elizabeth Graue invites us to her conversation with the authors of the previous four chapters by sharing her own perspective of various issues related to the discourse of readiness across fields. She problematizes forward-looking practices in that they function as exclusionary practices by removing "the actual child in the room" and replacing it with "an easy-to-teach student." She reminds us that this displacement of the here and now from education is not a new phenomenon resulting from the recent accountability and standard movement, but has a long history. Emphasizing the importance of the educator's agency and conscious practice, Graue explores "answerability" (Bakhtin, 1993) and "addressivity" (Bakhtin, 1986) as useful constructs that help the educator's ethical decision-making, responses, and interactions in schooling in sociohistorical contexts. Although critiqued in this book, Graue acknowledges development as "a key thread" in educational practice and suggests repurposing it through the idea of *developmentally responsive practice.* She argues that the notion of responsivity is critical to constructing educational practice that situates instruction within "moment-to-moment interactions" between teachers and pupils and that highlights the "contingent" nature of teaching and learning.

Together, the chapters in this section highlight how the construct of readiness has been reinforced across the fields. The authors in this section explore various alternative frameworks, such as Bakhtinian perspectives, Buddhism, and cultural psychology, which move beyond the linear and universal stages and age norms promoted by the last century's grand developmental theories (Damon, 1998; Lerner, 2006). I believe that the conversation around alternative theoretical perspectives among scholars across disciplines and fields will contribute to early childhood education's ongoing reconceptualization efforts by gaining insights into ideas and issues in other fields. I also hope that the cross-discipline and cross-field dialogue and collaboration may well lead to creating space for researchers exploring new frameworks in other fields, such as middle grades and teacher education, where debates over alternative perspectives have not been as active as they have been in the field of early childhood education.

REFERENCES

Bakhtin, M. M. (1986). *Speech genres and other late essays.* Austin: University of Texas Press.

———. (1990). *Art and answerability: Early philosophical essays* (M. Holquist & V. Liapunov, eds; V. Liapunov, trans.). Austin: University of Texas Press.

———. (1993). *Toward a philosophy of the act.* Austin: University of Texas Press.

Bruner, J. (1996). *The culture of education.* Cambridge, MA: Harvard University Press.

Cole, M. (1996). *Cultural psychology: A once and future discipline.* Cambridge, MA: Harvard University Press.

Damon, W. (1998). Preface to the handbook of child psychology, fifth edition. In W. Damon (series ed.) & R. M. Lerner (vol. ed.), *Handbook of child psychology: Vol. 1. Theoretical models of human development* (5th ed., pp. xi–xvii). New York: John Wiley & Sons.

Graue, M. E. (1992). Readiness, instruction, and learning to be a kindergartner. *Early Education and Development, 3,* 92–114.

———. (1993). *Ready for what? Constructing meanings of readiness for kindergarten.* New York: State University of New York Press.

Hanh, T. N. (1987a). *Being peace.* Berkeley, CA: Parallax.

———. (1987b). *The miracle of mindfulness.* Boston: Beacon.

———. (1993). *Interbeing.* Berkeley, CA: Parallax.

Kitayama, S., & Cohen, D. (Eds). (2007). *Handbook of cultural psychology.* New York: The Guildford Press.

Lerner, R. M. (1998). Theories of human development: Contemporary perspectives. In Damon & Lerner (eds), *Handbook of child psychology: Vol. 1* (5th ed., pp. 1–24).

———. (2006). Developmental science, developmental systems, and contemporary theories of human development. In W. Damon & R. M. Lerner (eds), *Handbook of child psychology: Vol. 1. Theoretical models of human development* (6th ed., pp. 1–16). New York: John Wiley & Sons.

Shweder, R. A., Goodnow, J., Hatano, G., LeVine, R. A., Markus, H., & Miller, P. (1998). The cultural psychology of development: One mind, many mentalities. In Damon & Lerner (eds), *Handbook of child psychology: Vol. 1* (6th ed., pp. 865–937).

CHAPTER 1

What's More Important: Numbers or Shoes? Readiness, Curriculum, and Nonsense in a Rural Preschool

Amy Noelle Parks and Sarah Bridges-Rhoads

One morning during circle time in her preschool classroom, Jakalah turned away from the counting book her teacher was reading to the class and began working on her untied shoes. As she became more engrossed in the task of tying, Jakalah stopped counting along with the class and neglected to respond to the prompts of the teacher. The classroom paraprofessional called Jakalah to the back of the room. "What's more important?" she asked Jakalah. "Numbers or shoes?" Without hesitation, Jakalah answered: "Shoes!" Seeing the paraprofessional's skeptical expression, she quickly amended her answer: "Numbers?" The paraprofessional agreed, saying: "That's right. Numbers. You need numbers for kindergarten."

This chapter, based on an analysis of incidents like this one, examines the ways in which nearly all of the interactions that occurred between teachers and children in a rural preschool classroom were framed by the discourse of "readiness" that has become so common in current conversations about early schooling (e.g., Lee & Ginsburg, 2007; Rothman, 2005; Stipek & Ryan, 1997). Recent literature in early childhood has been concerned with children's social and emotional readiness (DiBello & Neuhart-Pritchett, 2008), with ways of determining whether children are ready for kindergarten (Forget-Dubois et al., 2007), with interventions to be used on children who enter kindergarten *without* being ready (Winsler et al., 2008), and with gender differences in readiness (Guhn et al., 2007).

The curriculum used in Jakalah's preschool classroom, *SRA Language for Learning* (Englemann & Osborn, 1999), aimed to promote children's readiness for formal schooling through the direct teaching of what the authors call the "language of learning and instruction" (p. 4). The *SRA Language for Learning* teachers' guide tells teachers that:

> For many children, [language] instruction occurs informally in their homes and preschools before they reach kindergarten. But for other children, basic language instruction must occur in school. L for L offers these children this kind of instruction through carefully sequenced exercises that teach the concepts and skills they need to succeed in school. (Ibid)

Here, the authors explicitly state that the practices recommended in this curriculum are primarily for *some* children—those who do not learn language in their homes—and that the goal of the work is directed toward success at some later time, rather than at the day-to-day interactions of the preschool and kindergarten classroom.

We want to argue that many of the activities described in the curriculum and enacted in the classroom as part of this forward-looking language instruction could be described as nonsense. In using this word, we draw on a number of the dictionary definitions of nonsense, including *absurd, of little use, objectionable,* and, perhaps most importantly, *having little or no sense or meaning* (dictionary.com, 2008). In analyzing both the written and enacted curriculum in reference to these definitions, our goal is both to critique the particular pedagogy promoted in this curriculum and this classroom and to critique the readiness discourse in which this curriculum is located. At the end of the chapter we offer an alternative construct to readiness aimed at reframing the ethical and pedagogical responsibilities that teachers have in relation to their children.

CONTEXT AND METHODS

Before we proceed with the proposed analysis, we would like to describe a little bit about the school Jakalah attended. Oliver County Public School[1] is a PK-12 school with fewer than three hundred students in central Georgia. Nearly all of the school's students are African American, and as so many students qualify for free lunch, the school decided not to charge any of its students for meals. (These demographic characters may shed further light on the "some" children the *Language for Learning* curriculum proposes to teach.) This relatively new school was started as a public charter school in response to community concerns about children's academic performance in the school the next county over, where middle-school and high-school students had previously been bussed.

The European American teacher in Jakalah's preschool classroom (Mrs. Stanley) had been teaching in Oliver County throughout her twenty-plus-year career and had used a number of approaches to teaching young children including the current *Language for Learning* curriculum and whole language. The paraprofessional in the classroom (Mrs. Mason), who was African American, had been working in schools for nearly as long, and the pair had worked together in preschool for almost a decade. During the year of the observations, twenty students attended the preschool, including one who identified as white, one who identified as biracial, and eighteen who identified as black.

Each day, the students and teachers followed a similar routine, beginning the day with a morning meeting and then breakfast, followed by about ninety minutes devoted to the *Language for Learning* curriculum. During this time, the class began working as a whole group on the floor doing oral language exercises and finished by working in small groups on workbook pages. Mrs. Stanley and Mrs. Mason each led one of these groups, drawing on the *Language for Learning* teachers' guide as they gave directions to the children. The children then had about forty-five minutes of free-choice centers before lunch. After lunch, the children went out for recess, napped, ate a snack, and sometimes engaged in a short activity, such as watching a video, planting seeds, or doing an art project.

Our analysis is based on a six-month, interpretive ethnographic study (Dyson & Genishi, 2005; Emerson et al., 1995; Erickson, 1986). Amy visited the classroom once a week, took digital pictures, wrote field notes, interviewed children, and audio-taped classroom interactions. (Audio tapes were transcribed weekly.) Sarah analyzed the written curriculum. Although this chapter is based on data collected during a particular six-month period, we continued to work in the same preschool classroom the following year, and these experiences informed our thinking during analysis, which began with open coding (Emerson et al., 1995). Initially, we examined differences between spontaneous conversations in the classroom and conversations heavily shaped by the scripts in the *Language for Learning* curriculum. We then examined the scripted conversations in more detail, identifying particular instructional routines, some of which we will explore in depth in the following section. We then contrasted the opportunities for learning available in these routines with those available in the more spontaneous conversations. Finally, we interviewed both the children and the classroom teachers, asking about their opinions of the curriculum, and used these comments to add richness to our interpretations of the data.

CURRICULUM AND NONSENSE

In this section we will provide examples of three particular kinds of nonsense: the following of rules, the repeating of artificial sentences, and the coloring and interpretation of unusual pictures. The analysis will show that each of these practices worked together in the pre-kindergarten classroom to separate real life from school work, to discourage sense-making, and to reduce opportunities for meaningful conversations in genuine contexts.

Following Rules

One frequent instruction routine promoted in the *Language for Literacy* curriculum is learning and using rules for classification. The curriculum asks the children to learn the rules for ten classes, including:

- "If it's made to take you places, it's a vehicle."
- "If you can put things in it, it's a container."
- "If you can eat it, it is food."

In the classroom, the children did frequently recite each of these rules as well as others—both when they were called upon to do so and at other times. Once during workbook time, Mrs. Mason, following the directions of the curriculum, asked the children to color all the vehicles purple. She then asked:

> *Mrs. Mason*: What's the rule about vehicles?
> *Xavier*: If it's made to take people places, it's a vehicle.
> *Mrs. Mason*: No, I mean. What I just told you. What is the rule about what you are supposed to do with vehicles right now?
> *Xavier*: Color the vehicles purple?
> *Mrs. Mason*: Yes.

Later lessons asked children to apply these rules by assigning pictured objects to various classes, such as in one lesson when children had to decide whether pictured objects were food or tools. Mrs. Stanley held up a picture of a banana and asked Tameka if it was a tool. Tameka emphatically shook her head.

> *Tameka*: No.
> *Mrs. Stanley*: Say the whole thing, Tameka.
> *Tameka*: This is not no tool!
> *Mrs. Stanley*: This is not a tool. Say it with me.
> *Together*: This is not a tool.
> *Mrs. Stanley*: What is this, Tameka (pointing to the banana)?
> *Tameka looked up at the picture for a moment and then at Mrs. Stanley.*
> *Tameka*: A banana?
> *Mrs. Stanley*: And what is a banana?
> *Tameka opened her mouth and then closed it and looked at Mrs. Stanley.*
> *Mrs. Stanley*: It's a type of…
> *Harris shouted*: If you can eat it, then it's food.

In these episodes, both Tameka and Xavier struggled to participate in conversations where the goals were to attend to the rules of a complicated game rather than to use language to make sense of the world around them. Xavier got temporarily confused by which rule about vehicles he was supposed to produce at the particular time, while Tameka seemed puzzled by how she was supposed to reply to her teacher's question about the banana. We can understand Tameka's puzzlement if we remember that in Tameka's lived world, she knew that she and her teacher shared conceptions about what a banana was. They had eaten bananas for breakfast and had spoken about enjoying them. Tameka, who frequently played in the kitchen during center time, had occasionally served her teacher a plastic banana and had watched her "peel" and "eat" it. However, in the context of this curriculum-inspired conversation about rules, Tameka was supposed to forget about the shared knowledge she and her teacher had about this fruit and respond as if a "banana" was an abstract concept that existed only in the school world. Through this rule activity, the curriculum positioned Tameka as someone who must pretend that her teacher does not know what a banana is and that she herself is only able to identify it because she has memorized "the rule about food." The result is a conversation that is both absurd and lacking in sense. Over time, many children seemed to accept this difference between the lived world where they expected to make sense and negotiate and the school world presented by the curriculum.

For example, during one lesson, Tyronne was instructed to color all the containers red. He colored a cup and a bowl, leaving the table uncolored. He hesitated a moment before he began to color the vase red as well, but as he did so, he turned to me and whispered: "A vase ain't no container to put things in it." Here, Tyronne disputed the accuracy of the "rule" about containers he had been made to memorize, all the while coloring the vase red in recognition of the fact that his own opinion on the subject did not matter in this context.

In contrast to interactions like these, informal play during center time offered very different opportunities for language use. For instance, Tameka routinely used language about food while playing in the kitchen. One morning, Tameka and her friend Zadie proposed to make lunch for me. Tameka began by bringing me a bowl of spaghetti, saying: "You eat all that up now or you won't get no dessert." As I pretended to eat, she returned to the shelf. "You gotta have fruit." She came back with a bowl containing a plastic orange, banana, and a slice of watermelon. I took the orange and began to eat. Tameka took both bowls and dumped them in the sink where Zadie was standing. Zadie sighed and wiped her forehead, remarking: "Now, *I* gotta wash all these dishes. Get me that rag." In this short scene, Tameka demonstrated that she fully understood both the categories of food and fruit and used language to invite me into her pretend world. It is important to point out that this is a very different kind of pretend than the one in which the curriculum invited Tameka to engage. During centers, both Tameka and Zadie got to try out the roles of adult women, imagining themselves responsible for choosing a menu, enforcing dietary guidelines, and keeping the house clean—in addition to using oral language. *This* pretend play was neither absurd nor of little use.

Repeating Sentences

In the episodes described earlier, the children frequently used phrasing that deviated grammatically from formal English. The *Language for Learning* curriculum expressed concern about these deviations (which could probably be seen as markers for which children fell into the category of the "some" children who did not learn language at home). The curriculum encouraged teachers to correct children quickly and repeatedly if they did not "respond as the exercise is written" (Engleman & Osborn, 1999, p. 19) in the curriculum. In justifying the instructional routine of repeating sentences, the authors of the *Language for Learning* curriculum wrote: "Remember, the ability to accurately repeat statements the first time or after only a few practice trials is a good indicator of success in future academic work, including the ability to read and comprehend" (ibid.). To promote this repetition skill, which the authors described as essential for later learning in literacy, children were asked on a near daily basis to make statements in which they "say the whole thing," which means making a statement in a complete sentence and in a grammatically correct way, such as in the following episode.

> *Mrs. Stanley*: All right. Do the same thing that I'm doing. (She pointed at the ceiling and most children imitated her.) What am I doing?
> *Children*: Pointing at the ceiling! Pointing at the ceiling!
> *Mrs. Stanley*: Pointing at the ceiling. Say the whole thing. Say: We.
> *Children*: We.
> *Mrs. Stanley*: Pointed.
> *Children*: Pointed.
> *Mrs. Stanley*: At the ceiling.
> *Children*: At the ceiling.
> *Mrs. Stanley*: Say it all.
> *Children*: We pointed at the ceiling.

When engaged in this "say the whole thing" game, Mrs. Stanley stopped to correct the children both when they did not speak in complete sentences as in this example and when they used nonstandard English, as in the example with Tameka and the banana. These sorts of corrections were encouraged by the curriculum, which offered suggestions in nearly every lesson for how the teacher was to correct the children, most of which Mrs. Stanley ignored. The curriculum suggested that when correcting, teachers "model, lead, test, and re-test" (p. 25). For example, if some children were having difficulties saying the whole statement, "This is a cup," the teacher would be encouraged to respond as follows:

> *Teacher*: My turn. Listen. This is a cup. (Model)
>   Let's say it together. (Lead)
> *Teacher and children*: This is a cup
> *Teacher*: Your turn. Say the whole thing. (Test)
> *Children*: This is a cup.

The curriculum then recommends that the teacher return to an earlier part in the exercise where she would determine whether or not children were able to make the statement within the context of the lesson, thus completing the correction procedure with the retest, highlighting for the children that the various steps in the learning process "fit together in a sequence" (ibid.).

Oftentimes, during these oral language exercises children would latch on to some concept and begin calling out about it. For example, during one lesson, Mrs. Stanley broke away from the script to emphasize that plurals have an "sss" sound at the end, saying "sssss...like snake." Tyronne immediately responded to this, volunteering:

> *Tyronne*: Snake bite you. He bite your arm. (He gripped his arm to show how it would happen.)
> *Mrs. Stanley*: Mmm-hm. That's why we tell you not to play with snakes.
> *Helena*: Cause they'll bite you up.
> *Mrs. Stanley*: Are some snakes poisonous?
> *Children*: Yes! Yes!
> *Tyronne*: If you put him in a doghouse, he still might get out.
> *Cedric*: A snake can bite a dog and turn it into a snake.
> *Tameka*: They will bite you on the leg.
> *Amera*: They could bite your tongue!
> *Mrs. Stanley*: They could bite you anywhere.
> *Helena*: On the ground!
> *Zadie*: In water too!
> *Helena*: I don't like snakes.
> *Cedric*: I do. I do like snakes. I got—
> *Angel*: I seen those that are green and yellow.
> *Jerome*: I got one that brown...at my house.
> *Mrs. Stanley*: Okay, okay (*glancing at the clock*). Shhhh! We need to do our oral language. Touch your nose.

In this episode, the children eagerly share their opinions about snakes. Most use complete sentences and Cedric and Tyronne use sentence constructions that are more complicated than many phrases the curriculum suggested that children repeat. Initially, Mrs. Stanley encouraged this conversation, asking about whether snakes are poisonous and listening as the children spoke without correcting their grammar or asking them to repeat their sentences until they matched Standard English. Then, seemingly without irony, she noticed the time and silenced the children so they could practice their "oral language." It is easy to be critical of Mrs. Stanley in this moment where she stops the children's engagement in a meaningful conversation to pursue a scripted one; however, it is important to remember that she is situated in broader discourses. Her school has promoted a curriculum that is based on the idea that the language her children bring in from home is not good enough for them to be successful in their later school careers. As a preschool teacher, her job has been primarily defined as getting children ready for later schooling. In this discourse system, if she wants her children to be ready for kindergarten, then she must get them through this curriculum that proposes to teach them the language of schooling. However absurd or nonsensical the action of quieting children to practice language might appear from the outside, it makes perfect sense in a system where children are seen as not ready because of the ways they speak in their homes.

Interpreting Unusual Pictures

The final instructional routine we wish to highlight is the coloring and interpretation of unusual pictures. Much of the *Language for Learning* curriculum centers on children's talk about pictures in the teachers' guide and their coloring of pictures in the workbooks. Many of these pictures juxtapose images and settings in ways that make it difficult to tell coherent stories about the pictures and very few have any context, such as a written story, that might help children (or adults) to make sense of them. Two examples of these figures include a picture of an owl climbing a ladder and a picture of a bear in a wagon; both examples are discussed in this section.

Some of the pictures used for oral language showed relatively coherent scenes, such as girls eating ice cream cones. However, many times the pictures contained relatively inexplicable settings, such as a figure that shows an owl climbing a ladder, apparently to reach an apple. No explanation is given as to why the apple is on top of the ladder or why the owl climbs the ladder as opposed to flying. Instead, the point of the lesson is for children to practice speaking in complete sentences as they describe the picture. In the classroom, the lesson went like this:

Mrs. Stanley: Jerome, What did the owl do?
Jerome: He climbed up the ladder.
Mrs. Stanley: Tameka, what did he do after he climbed up the ladder?
Tameka: He ate the apple.
Mrs. Stanley: All right. Let's say it together. Say, first, the owl...
Children: First, the owl...
Mrs. Stanley: climbed the ladder.
Children: climbed the ladder.
Mrs. Stanley: After he climbed the ladder...
Children: After he climbed the ladder...
Mrs. Stanley: He ate the apple.
Children: He ate the apple.
Mrs. Stanley: Very good.

Here, when confronted with a picture that made little sense the children were not asked to create an explanation for it, to generate a story, or to discuss its absurdity. Instead, they were asked to engage in rote practice speaking. Occasionally, the curriculum, perhaps in order to promote higher order thinking, would invite teachers to ask children to comment on the picture during the oral language lesson. On these occasions children were frequently asked to describe something that was wrong with the picture. Often the children's perception of "wrong" and the curriculum's were quite different. For example, one picture portrayed a movie theater with two adults and three children sitting in seats. One child was reading a book with a flashlight. When asked what was wrong, several children said there weren't enough people in the theater for it to be a movie. Another child commented that the theater was not dark enough to show a movie. The curriculum's intended answer was that it was absurd that the child was reading in the movie theater. The curriculum prompted teachers to ask children about appropriate places to read. Thus, the possibility for critique that the curriculum opened to the children was quite limited, extending only to the improbabilities intended by the authors rather than to the ways in which the curriculum represented (or did not represent) the world as the children experienced it.

While the goals of the pictures in the teachers' guide were to promote daily language practice, the goals of the workbook pages were to provide opportunities for children to apply what they were learning to new contexts as well as to introduce some new concepts such as colors, shapes, matching, and picture-completion. Additional goals were to provide practice in following directions and in performing motor skills associated with classroom activities (Englemann & Osborn, 1999). To meet these goals, the curriculum (and in the observed classroom, the teacher) asked children to color in particular objects in the picture (such as food or containers), to cross other objects out, to draw lines between objects that had something in common, and to draw in missing objects. Sometimes these directions made a certain amount of intuitive sense, such as when students were asked to draw missing wheels onto a wagon. At other times, however, the directions lacked broader meaning. For example, children were often asked to color objects as a way of identifying them or classifying them, such as "Color all the food red," or "Color all the animals green." As a result, most children stopped thinking about what made sense in terms of coloring the picture. This was highlighted one morning when a substitute teacher took over one of the groups. She gave directions for the workbook page without looking at the teachers' guide. At one point she asked the children, "What color should the apple be?" All the children looked silently at her for a moment and finally Tameka said: "We don't know!" in exasperation. The substitute threw up her hands and said: "Well, what color are apples?" Xavier tentatively volunteered: "Red?" "Yes, red! Red. Color the apples!" she said. A few children picked up red crayons, but Angel, still unsure, said: "We color them red?" The children in this episode clearly did not expect to draw on their lived world knowledge of apples in order to make decisions about how to complete the workbook page. Experience had taught them that thinking about what made sense bore little relation to the marks they were expected to put on the page.

Often though, the children did attempt to impose meaning on sometimes inexplicable pictures. For example, when working on a picture from the workbook, Latisha turned to me and pointed at the page.

Latisha: Look, he's silly. A bear doesn't go in a wagon.
Amy: No, he doesn't.
Latisha: He look scared. He want to get out. He want to get out now. Mrs. Stanley, the bear want to get out!
Mrs. Stanley: He does look scared. Now (to the whole group) color the handle of the wagon green.

Here Mrs. Stanley briefly commented on the story Latisha had created about the bear, but then immediately redirected her to the curriculum's set task of coloring various parts of the picture. An opportunity to use language to construct a meaningful story was again pushed aside in favor of completing regimented tasks imposed by an outside source. Over time, the children grew quite savvy in their interpretation of the purpose of this curriculum. When I interviewed children, I asked them what they thought they were supposed to learn during their oral language and workbook time. Seven of the children immediately said "to follow directions," while ten said "coloring." Tyronne expanded on this when I spoke to him.

Amy: Can I ask you a question? When you do these workbook pages with Mrs. Stanley, what do you think you're supposed to learn?
Tyronne: Some. Some. Some. None. None. None. (He jabbed at the workbook page, which showed flowers with no petals, some petals, and many petals.)
Amy: Anything else?

Tyronne: Coloring. Color this, this and this. Color this red. Cross
   that out. Like that.
Amy: Why do you need to know that?
Tyronne: For kindergarten.

What is remarkable about the children's comments is how closely
they resemble the reasons for the curriculum given by the authors,
who justify the work in terms of what is needed for later schooling.
Both the curriculum and the children in the classroom described the
purpose of these schooling routines as teaching a particular set of
narrow skills that would be needed in the future. In fact, when I
asked a group of children painting what their favorite part of preschool was, three of them shouted out "kindergarten!" The teachers
in the classroom also promoted this forward-focused way of thinking,
frequently calling on kindergarten (as Mrs. Mason did in the opening
vignette) to justify a practice that may have seen absurd or meaningless in the moment. Although at Oliver County Public School children in preschool and kindergarten did not take the state's standardized
test, this assessment loomed over much of their work. Mrs. Stanley
said that she had some concerns about the curriculum, but acknowledged that because children in the primary grades had been making
the Annual Yearly Progress goals required under *No Child Left
Behind*, she thought it was unlikely that any changes in the preschool
curriculum would be approved. Thus, the students' anticipated performance on tests to be taken two years in the future shaped most of
the instructional interactions that occurred in the preschool.

OFFERING A CRITIQUE: BEYOND "IT'S NOT DEVELOPMENTALLY APPROPRIATE"

Many early childhood educators would question the educational
practices described in the previous section. Few would advocate the
repeating of sentences, the workbook pages devoid of greater meaning, or the reduction in opportunities for genuine communication.
However, some educators might frame their critique in terms of developmental appropriateness. That is, they would argue that the practices described earlier should not have been used with four-year-old
children in a preschool classroom because the children were not ready
for them or because the practices did not take into account the developmental characteristics of the children in the room. For example,
Bredekamp and Copple (1997) asserted that young children should
be in classrooms that provide many opportunities for active learning

and for play. They wrote: "When teachers provide a thematic organization for play; offer appropriate props, space, and time; and become involved in the play by extending and elaborating on children's ideas, children's language and literacy skills can be enhanced" (p. 14). This is clearly a different view of literacy learning than the one expressed by the authors of the *Language for Learning* curriculum and provides an opening for critique by calling attention to the ways that repeating sentences and coloring workbook pages reduced opportunities for play and for meaningful conversations with adults.

However, there are two major problems with situating the critique of this curriculum in the discourse of developmental appropriateness. The first problem is that saying the practices described earlier were not developmentally appropriate for four-year-olds implies there is some age or stage where coloring the vehicles purple or learning to repeat the rule for food becomes appropriate. It implies that these are educational practices that we would advocate for some children, at some time. However, it is difficult to think of an age level where the daily completion of workbook pages similar to those described earlier makes pedagogical sense. For this reason, the developmental perspective doesn't give enough leverage for the critique.

The second problem with situating the critique within developmentalism is more subtle and is based on the proposition that developmentalism itself contributes to a discourse of readiness that makes pedagogical practices such as those advocated by the *Language for Learning* curriculum and practiced in the observed preschool appear sensible. One of the key tenets in the developmental discourse is that "development occurs in a relatively orderly sequence, with later abilities, skills and knowledge building on those already acquired" (p. 10). Although educators are cautioned not to overemphasize age norms, bulleted lists of "widely held expectations" for various age levels convey the message that children of a particular age ought to be able to do particular tasks. If many children in a certain group do not achieve these expectations, then teachers may be likely to see this as a result of a group of children being behind and may feel that these children need special attention (or instructional routines) in order to meet these goals.

Bredekamp and Copple (1997) wrote that four-year-olds have vocabularies of four thousand–six thousand words, speak in five- to six-word sentences, and use advanced sentence structures. These expectations make it difficult for educators to see the four-year-olds in their classrooms except in relation to the standards. Developmental expectations (even with softening language about cultural and individual variations) make it possible to think of children as ahead or behind and frame teaching as progress. Moving to the next stage becomes the primary instructional goal irrespective of whether that movement requires physical, intellectual, or emotional development. In fact, the *Language for Learning* curriculum is designed in response to this concern and proposes extraordinary measures for the "some" children who fall short of the norm. The goal of the curriculum is to help these children meet the developmental expectations of a future grade level by doing intensive work now.

For example, the *Language for Learning* curriculum rationalizes many of its actions by referring to how it will make children "ready" for the types of experiences they will have in school. The curriculum states that it is not emphasizing language for "social communication," but in order to "transmit and receive important information, solve problems, and engage in higher-order thinking" (Engelmann & Osborn, 1999, p. 4). As a result, children spend a great deal of time drawing lines, crossing out drawings, and coloring pictures not because of the reading and learning they are doing now, but as necessities for future success in the following of directions that future curricula will give.

From this standpoint, developmentally appropriate instruction can occur once children have been made ready for it. In the observed preschool classroom, the justification offered by the teachers, the curriculum, and the students for many of the activities is that they will help children get ready for kindergarten. No one justified the instructional routines described earlier as benefiting children in the moment. This is the danger of challenging the *Language for Learning* curriculum primarily in terms of developmental appropriateness; it fails to disrupt the focus on forward-thinking. As a result, routines that many find objectionable are seen as acceptable pedagogical moves because they are seen as leading to some greater good in the future. In the next section we propose a way of thinking about the work of teaching preschool that does not depend primarily on either readiness or developmental appropriateness as a justification for action.

AN ANSWERABLE EDUCATION

In the previous section, we argued that critiques that primarily ask "Is this pedagogical practice *developmentally appropriate?*" fall short of disrupting instructional practices such as those emphasized in *Language for Learning*. We would like to propose the alternative question of "Is this pedagogical practice *answerable?*" as an ethical interrogation of teaching. The concept of answerability comes from the work of Russian literary critic Bakhtin. In discussing answerability, Bakhtin (1990) emphasized each human being's ethical responsibility to the moment and to the people with whom he or she shares that moment. He wrote: "I myself—as the one who is actually thinking and who is answerable for his act of thinking—I am not present in the theoretically valid judgment. The theoretically valid judgment, in all of its constituent moments, is impervious to my individually answerable self-activity" (p. 4). Here Bakhtin, in response to Kantian notions about categorical imperatives, articulated an ethics based on an individual's responsiveness to the particularities of any given moment. In contrast to Kant who argued that ethical judgments ought to be formed in abstract and as though the individual knew none of the people or contexts involved, Bakhtin argued that people ought not create theoretical rules or standards ahead of time and apply them in a universal way to particular moments in our lives; nor, should individuals project ahead to the future as justification for actions in the current moment. For example, Kant would probably agree that a description of good teaching or a good early literacy lesson could be developed and applied across contexts while Bakhtin's writing would suggest that particular contexts and meanings would make such an endeavor impossible.

In drawing on Bakhtin's work, we argue that educators must work to create a discourse where teachers are expected to be answerable in the moment to the students in the room with them, rather than to justify their actions in relation to "theoretically valid" standards or the threat of the future. In articulating this ethics of answerability, Bocharov (1993), a Bakhtinian scholar, wrote that: "A human being has not a right to an alibi—to an evasion of that unique answerability which is constituted by his actualization of his own unique, never repeatable 'place' in Being: he has no right to an evasion of that once-occurrent 'answerable act or deed' " (p. xxii). This lack of an alibi means that human beings cannot evade their ethical responsibilities in any moment. We live each of these moments only once and in each we must exercise our own unique ethical choice in relation to other human beings.

For educators, the state standards, high-stakes testing, and the opinions of teachers in future grade levels cannot, do not, serve as alibis for the responsibility teachers have for the students in front of them at any moment. Rather than justifying pedagogical choices by drawing on these outside standards or norms or by pointing toward the future, an ethics of answerability would require educators to justify choices in relation to the small human beings who share each unique moment with them. In terms of the *Language for Learning* curriculum, this means that practices such as saying the whole thing, reciting rules, and interpreting ridiculous pictures would have to be justified in terms of their power to engage, enchant, or enrich the four-year-olds in the preschool classroom in the moments they enact these practices. We would argue that the practices described here, which encouraged students to separate their work on school tasks from their meaning-making in their broader lives, were not answerable to the children in the preschool classroom and therefore were unethical.

Does this mean that educators must ignore the very real sword that hangs over many children's heads as they march toward the standardized-testing grade levels? Absolutely not. It does mean though that concern for children's futures cannot outweigh responsibility—or answerability—to their present. It means that questions about a pedagogical practice's answerability must come prior to questions about its efficacy. It means that educators must create a teaching discourse that encourages preschool teachers to care about whether children can walk in the shoes they have on their feet before they worry about whether children can recite the expected numbers in kindergarten.

Hicks (1996), in discussing Bakhtin's ethics, wrote that the commitment required by answerability was "more similar to faithfulness, even love, than to adherence to a set of norms" (p. 107). Taking this active, conscious stance of faithfulness toward children requires that educators come to each moment with an awareness of the present. As educators worry about whether children will be ready for kindergarten or for testing, they must filter that worry through a love of the children in the here and now. If a practice cannot be justified in each "unique and never-repeatable place," then an alternative must be found. When teaching answerably, concern for the future is valid, but cannot supersede the present. Faithfulness to children in the moment calls into question both scripted curricula and developmental standards. Instead, an answerable pedagogy demands that teachers come to children in each moment with serious attention to their present selves, rather than with somebody else's nonsense.

NOTES

The research presented in this chapter was funded in part by a grant from The Spencer Foundation.

1. The names of the school, the teachers, and the children are all pseudonyms.

REFERENCES

Bakhtin, M. M. (1990). *Art and answerability: Early philosophical essays* (M. Holquist & V. Liapunov, eds; V. Liapunov, trans.). Austin: University of Texas Press.

Bocharov, S. G. (1993). Introduction to the Russian edition. In M. M. Bakhtin, *Toward a philosophy of the act* (M. Holquist & V. Liapunov, eds; V. Liapunov, trans.; pp. xxi–xxiv). Austin: University of Texas Press.

Bredekamp, S., & Copple, S. (1997). *Developmentally appropriate practice in early childhood programs* (revised ed.). Washington, DC: National Association for the Education of Young Children.

DiBello, L. C., & Neuhart-Pritchett, S. (2008). Perspectives on school readiness and pre-kindergarten programs: An introduction. *Childhood Education, 84*(5), 256–259.

Dictionary.com Unabridged (v 1.1). nonsense. (n.d.). Retrieved September 25, 2008, from Dictionary.com website: http://dictionary.reference. com/browse/nonsense.

Dyson, A. H., & Genishi, C. (2005). *On the case: Approaches to language and literacy research.* New York: Teachers College Press.

Emerson, R. M., Fretz, R. I., & Shaw, L. L. (1995). *Writing ethnographic fieldnotes.* Chicago: University of Chicago Press.

Engelmann, S., & Osborn, J. (1999). *Language for learning: Teacher's guide.* Worthington, OH: SRA/Mcgraw-Hill.

Erickson, F. (1986). Qualitative methods in research on teaching. In M. C. Wittrock (ed.), *Handbook of research on teaching* (pp. 119–160). New York: Macmillan.

Forget-Dubois, N., Lemelin, J., Boivin, M., Dionne, G., Seguiin, J. R., Vitaro, F., & Tremblay, R. E. (2007). Predicting early school achievement with the EDI: A longitudinal population-based Study. *Early Education and Development, 18*(3), 405–426.

Guhn, M., Gaderman, A., & Zumbo, B. D. (2007). Does the EDI measure school readiness in the same way across different groups of children? *Early Education and Development, 18*(3), 453–472.

Hicks, D. (1996). Learning as a prosaic act. *Mind, Culture & Activity, 3*(2), 102–118.

Lee, J. S., & Ginsburg, H. P. (2007). Preschool teachers' beliefs about appropriate early literacy and mathematics education for low- and middle-socioeconomic status children. *Early Education and Development, 18*(1), 111–143.

Rothman, R. (2005). Testing goes to preschool: Will state and federal testing programs advance the goal of readiness for all children? *Harvard Education Letter, 21*(2), 1–4.

Stipek, D., & Ryan, R. (1997). Economically disadvantaged preschools: Ready to learn but further to go. *Developmental Psychology, 33*, 711–723.

Winsler, A., Tran, H., Hartman, S. C., Madigan, A. L., Manfra, L., & Bleiker, C. (2008). School readiness gains made by ethnically diverse children in poverty attending center-based childcare and public school prekindergarten programs. *Early Childhood Research Quarterly, 23*(3), 314–329.

CHAPTER 2

Who is Normal? Who is Abnormal? Rethinking Child Development from a Cultural Psychological Perspective

Kyunghwa Lee

Throughout the twentieth century, developmental psychology has greatly influenced the field of early childhood education (hereafter, the field). Educators have used developmental theories to talk about the norms for children's development and learning. Although since the early 1990s some early childhood educators (e.g., Bloch, 1992; Cannella, 1997; Jipson, 1991; Kessler & Swadener, 1992; Mallory & New, 1994) have begun to critique the field's reliance on developmental psychology, many policies and practices related to curriculum and educational experiences provided for young children are still based on the past century's developmental perspectives.

Some educators have eloquently pointed out that developmental theories alone cannot answer all questions about education for young children (e.g., Spodek, 1988). Some reconceptualists have questioned whether the knowledge of child development is worthy of early childhood educators' attention (e.g., Cannella, 1997; Ryan & Grieshaber, 2005). Although I agree with these educators' concern about the dominance of developmental psychology as the knowledge base, my critique of the field's focus on developmental theories takes a different tack. Instead of rejecting insights about human development, I problematize that the field has maintained an allegiance to an outdated and limited developmental perspective, a perspective no longer viable even in the field of developmental psychology.

For this purpose, I, first, provide a brief overview of developmental theories, including contemporary developmental perspectives, and their influence on early childhood education. Next, I draw on data from my research (Lee, 2008) on early childhood teachers' perceptions about children with attention deficit/hyperactivity disorder (ADHD) and illustrate how an outdated and limited developmental perspective based on the notions of developmental stages and readiness is still used to identify abnormality in children's behavior. I then introduce cultural psychology (e.g., Bruner, 1996; Cole, 1996; Kitayama & Cohen, 2007; Shweder et al., 1998) as a contemporary developmental systems model (Lerner, 1998, 2006) and discuss how this alternative framework might help early childhood teachers move beyond the limitations of the past century's developmental theories by understanding the importance of context in their perceptions about children's development, including disability.

Developmental Theories and Early Childhood Education

The dominant developmental perspective adopted by the field reflects the "three grand systems" (Damon, 1998, p. xv) of the twentieth century, which included Piaget, psychoanalysis, and learning theory. These grand theories described development as universal and identical to everyone across time and place. These theories also perceived development as an individualistic process and focused on biology and evolution (Bruner, 1996).

By the end of the twentieth century, however, the viability of the grand developmental theories had considerably weakened in the field of developmental psychology. In his analysis of the successive editions of the *Handbook of Child Psychology* published in 1946, 1954, 1970, and 1983, Damon wrote, "The grand old theories were breaking down. Piaget was still represented by his 1970 piece, but his influence was on the wane throughout the other chapters. Learning theory and psychoanalysis were scarcely mentioned" (1998, p. xv). The fifth edition of the *Handbook* published in 1998 noted that the grand developmental theories had been replaced by "a systems perspective" (Lerner, 1998). Lerner explained this contemporary trend as follows:

[The] mechanistic and atomistic views of the past have been replaced by theoretical models that stress the dynamic synthesis of multiple layers of analysis...In other words, development, understood as a property of systemic change in the multiple and integrated levels of organization (ranging from biology to culture and history) comprising human life and its ecology, or, in other words, a *developmental systems perspective*, is an overarching conceptual frame associated with contemporary theoretical models in the field of human development. (P. 2; emphasis in original)

Unlike the grand developmental theorists, contemporary developmentalists are interested not in "structure, function, or content per se, but in change, in the processes through which change occurs, and thus in the means through which structures transform and functions evolve over the course of human life" (p. 1). Lerner further elaborated:

A developmental systems perspective involves the study of active people providing a source, across the life span, of their individual developmental trajectories; this development occurs through the dynamic interactions people experience with the specific characteristics of the changing contexts within which they are embedded. (P. 16)

The most recent volume of the *Handbook* published in 2006 reiterates the importance of understanding the complexity of human development by using systems models. As an interesting new trend, Lerner (2006) explained that scholars interested in human development are no longer limited to developmental psychologists. Increasing multidisciplinary approaches to human development have led scholars in this area to refer to their field not as developmental psychology, but as "developmental science" (p. 4) in order to be comprehensive.

Most importantly, contemporary developmentalists utilizing the systems models have appreciated postmodern ideas and have avoided the long-standing Cartesian dichotomies, such as nature-nurture and biology-culture, to highlight that "genes, cells, tissues, organs, whole organisms, and all other, extraorganism levels of organization composing the ecology of human development are fused in a fully coacting, mutually influential, and therefore dynamic system" (p. 10). Thus, a developmental systems approach emphasizes the "bidirectional relation" between the individual and context. Lerner argued that a developmental systems approach reveals a paradigm shift in developmentalism. He summarized contemporary approaches to human development as follows:

[Within] the context of the relational metatheory that has served as a product and a producer of developmental systems thinking (Lerner, 2002), there has been a rejection of an idea that is derived from the positivist and reductionist notion that the universe is uniform and permanent—that the study of human behavior should be aimed at identifying nomothetic laws that pertain to the generic human being. This idea was replaced by a stress on the individual, on the importance of attempting to identify both differential and potentially idiographic laws as involved in the course of human life...Similarly, the changed philosophical grounding of the field has altered developmental science from a field that enacted research as if time and place were irrelevant to the existence and operation of laws of behavioral development to a field that has sought to identify the role of contextual embeddedness and temporality in shaping the developmental trajectories of diverse individuals and groups. (P. 6)

This new trend among contemporary developmentalists, however, has had little influence on the field, as educators still discuss children's development along "conceptually implausible and empirically counterfactual lines" (Lerner, 1998, p. 2), still adhering to the notion of universal developmental stages and the false dichotomies of nature-nurture and biology-culture. For many educators, the focus remains on separate content domains, such as language development, cognitive development, motor development, social-emotional development, and so on.

In relation to the field's dependence on the old grand developmental theories, Walsh (1991) argued that in the United States the discussion on child development continues to rely heavily on Piagetian perspectives combined with the dominant romantic maturationism of the first half of the twentieth century. These perspectives have provided the foundation for the field's discourse of developmentally appropriate practice (Bredekamp, 1987; Bredekamp & Copple, 1997; Copple & Bredekamp, 2009) and have kept educators focusing on the notions of developmental stages and readiness. In the next section, I illustrate how these notions, stemming from the paradigm of the past century's grand developmental theories, are central to the discourse and practice for diagnosis and treatment of children with ADHD.

Developmental Norms, Readiness, and Disability

In an ethnographic study conducted with four Midwestern early childhood teachers,[1] I found that the teachers' folk theories about teaching and learning centered around identifying each child's areas of weakness and, often less frequently, strengths to provide individualized education. Particularly, all participant teachers expressed their concern about children with ADHD. I soon learned that diagnosis and medication treatment for ADHD was of nationwide interest in the United States. This topic made frequent appearances in the popular press, the media, academic publications, government reports (e.g., National Institutes of Health Consensus Statement, 1998), and parent magazines. For example, in a *Time* magazine cover story titled *The age of Ritalin,* Nancy Gibbs (1998) reported, "Production of Ritalin [medication used for ADHD] has increased more than sevenfold in the past eight years and 90% of it is consumed in the U.S." (p. 89). As a former kindergarten teacher in Korea, I found U.S. early childhood teachers' concern about ADHD and medication treatment foreign, and yet, fascinating. This interest provided the impetus for the cross-national study on early childhood teachers' perceptions of children with ADHD that I conducted with researchers in Denmark, England, Iceland, Japan, and Korea (Einarsdóttir et al., 2005; Lee & Neuharth-Pritchett, 2008). As part of this cross-national study, I invited ten Southeastern teachers for the U.S. portion of data collection (Lee, 2008). In this section, I draw on the data collected with these teachers.

ADHD at the National Context

Before presenting the analysis of data, I would like to provide brief contextual information about the discourse on ADHD at the national level generated by two powerful professional organizations: the American Psychiatric Association (APA) and the National Institutes of Health (NIH).

According to the NIH (1998) consensus statement about ADHD, this is "the most commonly diagnosed behavioral disorder of childhood, estimated to affect 3 to 5 percent of school-age children" (p. 5) in the United States. Lack of consistent clinical diagnosis (e.g., laboratory tests, neurological examinations) and "the inherent ambiguity of ADHD symptoms" (Singh, 2002b, p. 360), however, challenge the accurate identification of children with ADHD. To date, ADHD diagnoses heavily rely on, often subjective, observation data provided by families, teachers, and health professionals. Table 2.1 shows the widely used diagnostic criteria for ADHD presented in the most recent edition of *Diagnostic and Statistical Manual of Mental Disorders* (DSM) published by the APA (2000). The DSM defines three subtypes of ADHD: a combined hyperactive and inattentive type, a primarily inattentive type, and a primarily hyperactive-impulsive type (APA, 2000). As shown in table 2.1, the DSM criteria note that, for diagnosis, symptoms should persist "for at least 6 months to a

Table 2.1 Diagnostic criteria for attention-deficit/hyperactivity disorder

A. Either (1) or (2):
   (1) Six (or more) of the following symptoms of *inattention* have persisted for at least six months to a degree that is maladaptive and inconsistent with developmental level:

   Inattention
   (a) often fails to give close attention to details or makes careless mistakes in schoolwork, work, or other activities;
   (b) often has difficulty sustaining attention in tasks or play activities;
   (c) often does not seem to listen when spoken to directly;
   (d) often does not follow through on instructions and fails to finish schoolwork, chores, or duties in the workplace (not due to oppositional behavior or failure to understand instructions);
   (e) often has difficulty organizing tasks and activities;
   (f) often avoids, dislikes, or is reluctant to engage in tasks that require sustained mental effort (such as schoolwork or homework);
   (g) often loses things necessary for tasks or activities (e.g., toys, school assignments, pencils, books, or tools);
   (h) is often easily distracted by extraneous stimuli;
   (i) is often forgetful in daily activities.

   (2) Six (or more) of the following symptoms of *hyperactivity-impulsivity* have persisted for at least six months to a degree that is maladaptive and inconsistent with developmental level:

   Hyperactivity
   (a) often fidgets with hands or feet or squirms in seat;
   (b) often leaves seat in classroom or in other situations in which remaining seated is expected;
   (c) often runs about or climbs excessively in situations in which it is inappropriate (in adolescents or adults, may be limited to subjective feelings of restlessness);
   (d) often has difficulty playing or engaging in leisure activities quietly;
   (e) is often "on the go" or often acts as if "driven by a motor";
   (f) often talks excessively;

   Impulsivity
   (a) often blurts out answers before questions have been completed;
   (b) often has difficulty awaiting turn;
   (c) often interrupts or intrudes on others (e.g., butts into conversations or games).

B. Some hyperactive-impulsive or inattentive symptoms that caused impairment were present before age seven.
C. Some impairment from the symptoms is present in two or more settings (e.g., at school [or work] and at home).
D. There must be clear evidence of clinically significant impairment in social, academic, or occupational functioning.
E. The symptoms do not occur exclusively during the course of a pervasive developmental disorder, schizophrenia, or other psychotic disorder and are not better accounted for by another mental disorder (e.g., mood disorder, anxiety disorder, dissociative disorder, or a personality disorder).

American Psychiatric Association, 2000, pp. 92–93.

degree that is maladaptive and inconsistent with developmental level" (p. 92). Also, the criteria state that some symptoms and impairment are present "in two or more settings (e.g., at school [or work] and at home)" and "before age 7" (p. 92).

These diagnostic criteria describing the appearance of symptoms in two settings before age seven and many features related to difficulties in structured activities and schoolwork often lead early childhood teachers to be involved in the challenging process of ADHD diagnosis among children under their care. The DSM states:

> Symptoms of [ADHD] are typically at their most prominent during the elementary grades. As children mature, symptoms usually become less conspicuous. By late childhood and early adolescence, signs of excessive gross motor activity...are less common, and hyperactivity symptoms may be confined to fidgetiness or an inner feeling of jitteriness or restlessness. (P. 89)

Under these guidelines emphasizing the high visibility of ADHD symptoms among young children, oftentimes it is the early childhood teacher who first notices a difference in a child's behavior and initiates referral.[2] Because many features of ADHD are similar to behavioral characteristics of healthy children, however, differentiating between normal and abnormal behavior is a daunting task for teachers. The DSM acknowledged this difficulty, stating, "In early childhood, it may be difficult to distinguish symptoms of [ADHD] from *age-appropriate behaviors in active children* (e.g., running around or being noisy)" (p. 91; emphasis in original). Against this backdrop of the challenge of ADHD diagnosis faced by early childhood teachers, I present the ten teachers' discussion about children with ADHD.

ADHD at the Local Context

The ten teachers invited for the U.S. portion of data collection taught in pre-kindergarten through third grade classrooms (two teachers per grade) at two elementary schools in two different school districts in a Southeastern state. Each teacher participated in a semi-structured individual interview for about one hour. All participant teachers were European-American and, except for one kindergarten teacher, all were female. At the time of data collection, the teachers had four–thirty years of teaching experience (see Lee, 2008, for detailed information about research methods).

When asked about their experience with children with ADHD, all ten teachers shared that they had one–five children either already diagnosed or likely to be diagnosed as having ADHD each year. Unlike teachers in other nations such as Denmark, Japan, and Korea (Einarsdóttir et al., 2005; Holst, 2008; Hong, 2008), diagnosis and medication treatment for ADHD was a familiar discourse and practice to these U.S. teachers, and none of them questioned or expressed difficulty in accepting the notion of ADHD. Nonetheless, all ten teachers readily acknowledged the challenge of distinguishing ADHD symptoms from typical behavior of young children. This difficulty was more pronounced in the pre-kindergarten and kindergarten teachers than in their elementary grade colleagues as stated by Laura, a pre-kindergarten teacher: "It's harder with 4s, because we don't have seat work. We don't have homework, you know...And it's kind of characteristic of 4s to lack self-control."

As discussed earlier, an accurate diagnosis of ADHD is an area of concern even among health professionals. In acknowledgment of the controversy over the diagnosis and pharmaceutical treatment of ADHD, the NIH held a Consensus Development Conference in 1998. At the end of the conference, the panel concluded, "A more consistent set of diagnostic procedures and practice guidelines is of utmost importance" (NIH consensus statement, 1998, p. 3). To avoid or minimize misdiagnosis, the DSM emphasizes that the behavioral features indicative of ADHD should be "more frequently displayed and more severe than [what is] typically observed in individuals at a comparable level of development" (APA, 2000, p. 85).

The APA's emphasis (2000) on the identification of atypical behavior "inconsistent with developmental level" (p. 92) assumes that everyone involved in the process of ADHD diagnosis knows the norm of behavior at each developmental stage applicable to all children across time and place. A kindergarten teacher, however, shared his conundrum around this notion of the universal developmental norm:

> And especially, being at that age...in kindergarten, [it], I think, has to be an *extreme* case for it to be a very cut-and-dried, a clear thing at that age..., and it's still very challenging for me to be able to really see if it is an acceptable norm...for them to be in, or they're really beyond where they should be...as far as the norm.

In this excerpt, Ron expressed his difficulty in defining the range of normality for children's behavior because "an acceptable norm" is defined not by absolute criteria or rules, but by expectations and meanings created by people around children.

The absence of definite criteria for the norm encouraged the teachers to constantly compare a child's behavior with his or her age and grade peers in order to identify and justify abnormality in that child's behavior. Jennifer, a second grade teacher, said, "I think the ones that are ADHD are just over and above...They just seem to kind of take things a little too far. Again, constantly, constantly need to be redirected." Similarly, Michelle, another second grade teacher, elaborated on what made her convinced that a child has ADHD:

> I've noticed that sometimes little boys, especially in second grade—and in kindergarten and Pre-K—love to roll around and play with each other, and a lot of the ADHD kids have that tendency, but keep going. They...take it one step further. So the ones that have that activity level—say, a second grader with it that's *not* diagnosed ADHD—you can bring him back quicker than you can with an ADHD kid. Usually they [ADHD kids] take *more* cues, *more* asking to come back.

As features symptomatic of ADHD, both teachers somewhat ambiguously described behavior that takes situations "a little too far" or "one step further" and that requires more frequent reminders than what is observable from other children.

The comparison between children's behavior also included the length of time necessary for each child's adjustment to routines and expectations in school. Deb, a pre-kindergarten teacher, explained when she might begin noticing an abnormality in a child's behavior:

> Once they've been given time, 4 to 6 weeks [in the classroom], I think, max, and they're able to—I hate to say "conform," but that's really what it is—to the norms and the routines and the patterns and expectations, then that's one thing. And some of these children may still be active and busy, but they've learned how to control themselves, okay? Children who have been given those same kinds of opportunities and are not able to do that, and there are some students that are not able to do that...perhaps need to be evaluated.

According to this view, a child who takes longer than his or her peers to comply with classroom routines and expectations might be subjected to a referral. Interestingly, compared to the DSM that emphasizes the persistence of symptoms and impairment for "at least 6 months" for diagnosis, the criteria for potential ADHD held by this teacher have much shorter persistent periods (i.e., a maximum of four–six weeks after school starts for pre-kindergartners).

In relation to seeing the difficulty in adjustment to routines and expectations in school as a symptom of ADHD, four of the ten teachers thought that children with ADHD tended to be chronologically younger than their peers as shared by Janice, a third grade teacher: "A lot of times I do think you see it in your younger kids, because they just aren't...where the other kids are emotionally yet, and that factors into their attention spans." Related to this issue of age, Susan, a first grade teacher, shared her strong belief in the problem of children with late birthdays in school:

> I've read a lot about age differences, and how to fix children's performance in school, as far as...younger or older 6-year-olds when they start school. And there seems to be really a lot of evidence to support that the later birthday children...do have...more academic struggles and more attention struggles. Not necessarily in the first grade or kindergarten, but it might even show up in later elementary or middle school years, and so I'm a big proponent for children not starting school until they're *ready* to start school, and this comes from personal experience. One of my sons has a birthday in October, and so he...was about almost 6 before he started kindergarten, and he was such a good student and just did so well. Then when my second child came along, who had a July birthday, I purposefully kept him back so that he would be the oldest in his [classroom]...And also, my years of teaching [in] school made me realize that the older the child, the more successful they seem to be.

Whereas these teachers believed that immaturity resulting from late birthdays was a factor to be considered for ADHD diagnosis, two other teachers thought immaturity was not necessarily connected to age. In the following excerpt, a second grade teacher Michelle shared her opinion:

> [Children with ADHD] act very much more immature...Silly. Sometimes baby talk. Sometimes [they] will go into their shell really quickly, because they get mad...and frustrated, because they can't do...some work...Not age-wise...But in general,...they *act* younger.

These teachers' discussion about maturity reveals that their perceptions about children with ADHD are closely connected to the notion of "readiness." In her study on the meanings of kindergarten readiness, Graue (1993) discussed that "youngness" is seen as problematic, and that decisions about grade retention and referral to special education might "well be biased" because of the teacher's "belief about the relationship of child age and performance" (p. 9). Data from my study support that, whether chronologically young or acting young, the teachers perceived youngness in a negative light. More specifically, the teachers' discourse about ADHD shows that being young and not ready for school or classroom expectations can be considered pathological.

The ten teachers' descriptions of problematic behavior and ADHD largely overlapped as they focused on disruptive behavior (e.g., "high maintenance" children) and inattentive behavior (e.g., children being "off in [their] own world"). In these teachers' opinion, "anything that disrupts the regular flow of teaching" (Ron, a kindergarten teacher) was a serious problem in the era of the *No Child Left Behind (NCLB) Act*. This perception was particularly evident among upper elementary grade teachers as shared by Kim, a third grade teacher:

> [The problem behavior] just causes such distraction in your room that you're not able to teach your children. And, especially in third grade, their learning is so vital because there's so much pressure with the testing that we have to take. There's not a lot of time to deal with severe problems. You want everything to flow smoothly so everyone can learn the best that they can and get ready for everything that they need to.

According to these teachers, disruptive, inattentive, and immature behavior was a "distraction" that compromised their instruction time and that "[interfered] not only with that child's learning, but with...the learning of the other children in the class" (Susan, a first grade teacher).

These teachers' discussion indicates that their perceptions about problematic and atypical behavior were influenced by the pressure of accountability. Although, as described earlier, all ten teachers had difficulty in distinguishing ADHD symptoms from typical behavioral characteristics of young children, atypical behavior was eventually defined as the one that negatively affected children's "education," "grade," and "[promotion] to the next grade level." Two elementary classroom teachers shared that framing behavioral issues in the context of the child's mastery of academic skills and grade promotion is most effective when communicating with parents because this strategy helps minimize dissent between the teacher and parents in their perceptions about abnormality in a child's behavior. A first grade teacher shared:

> I like to just explain to [parents] that the behavioral issues are a problem because they are keeping this child from mastering first-grade skills,

and by the end of the year the child may not be where he needs to be or where she needs to be. I don't like to tell [parents], "This is a problem for me, in my classroom." I like to say, "This is a problem because *your* child is not moving forward at the rate he or she should move, and we know that [he or she is] capable of doing better, so let's figure out how we can help your child get to where he [or she] needs to be,…and master the material that needs to be mastered, and be able to test out at the end of the year and move onto the next grade level. We've become such a test-oriented school and society, that parents understand the importance for the children to show mastery on…standardized tests, and we know there will come a time when this will determine whether or not a child progresses to the next grade level.

This excerpt reveals that, in today's high-stakes testing environment, both teachers and parents are subjected to accept the diagnosis and treatment for ADHD in order to help children to be successful on standardized tests and to be ready for the promotion to the next grade level. In this context, all ten teachers, albeit cautious to prescribe medication for young children, rationalized their recommendation for the medication treatment for children with ADHD because they believed that children on medication "are able to function in a classroom much more effectively" (Susan, a first grade teacher) and that medicated children are "being successful in school" (Rebecca, a first grade teacher).

DISABILITY AND DEVELOPMENT FROM A CULTURAL PSYCHOLOGICAL PERSPECTIVE

As discussed earlier, contemporary developmentalists emphasize understanding human development by using the systems models that explicate the interplay between the individual and context (Lerner, 1998, 2006). In this section, I introduce cultural psychology (e.g., Bruner, 1996; Cole, 1996; Kitayama & Cohen, 2007; Shweder et al., 1998) as a systems model that might help early childhood educators move beyond the limitations of the past century's grand developmental theories that are still dominant in their discussion about children's learning and development. I begin the section with a brief overview of cultural psychology and then apply this framework to examine the discourse on ADHD.

Overview of Cultural Psychology

Cole (1996) traced the history of cultural psychology back to Wilhelm Wundt who opened his laboratory in Leipzig, Germany, in 1879 and introduced psychology as a discipline. According to Cole, Wundt imagined two psychologies: the first, experimental, and the second, cultural. Nisbett (2007) articulated Wundt's thought that "one could not understand behavior by just looking at what people were doing in laboratories. One also had to know history and culture" (p. 837). The first psychology, however, became dominant, and psychologists ignored Wundt's second psychology and neglected the role of culture (Bruner, 1996). It was not until the 1980s that a critical mass of scholars in various disciplines (e.g., activity theorists, anthropologists, contextual psychologists of the sociohistorical school, and some cognitive and developmental psychologists) began to show interest in how culture enters into the process of human development (Shweder et al., 1998). Shweder and his colleagues pointed out that cultural psychology is not a new field, but a revived one. Cole, in the subtitle of his book, referred to cultural psychology as "a once and future discipline."

To cultural psychologists, culture is the most significant system within which human development occurs. They consider culture "loosely coherent, collective interpretive frameworks that function as a shared moral narrative" (Lee & Walsh, 2001, p. 79). In order to understand culture, they believe careful attention needs to be paid to "the custom complex" (Shweder et al., 1998) by focusing on both what people do and what they think. Shweder et al. defined cultural psychology as

> the study of the way culture, community, and psyche make each other up. Alternatively stated, cultural psychology is the study of all the things members of different communities *think* (know, want, feel, value) and *do* by virtue of being the kinds of beings who are the beneficiaries, guardians, and active perpetuators of a particular culture. (P. 867; emphasis in the original)

By definition, cultural psychologists seek to understand "the co-creation of person and culture, [emphasizing that persons and cultures are] interdependent and mutually active" (Miller, 1998, p. 1). Highlighting this interdependence, Bruner (1996) noted that "culture though itself man-made, both forms and makes possible the workings of a distinctively human mind" (p. 4). Cultural psychologists distinguish themselves from cross-cultural psychologists who treat culture as an independent variable or an overlay on some universal process (Greenfield, 1997). Cultural psychologists do not assume that development is essentially universal and marked by minor cultural variations. Regarding cultural psychologists' perspective of the relation between culture and development, Shweder et al. (1998) elaborated:

> [It] is evident that culture does not surround or cover the "universal" child. Rather, culture is necessary for development—it completes the child. Culture provides the script for "how to be" and for how to participate as a member in good standing in one's cultural community and in particular social contexts. Simultaneously, a cultural psychology perspective recognizes that children are active constituents of their own cultures and that changes in individuals initiate changes in their relations with others and thus in their immediate cultural settings. (P. 896)

From a cultural psychological perspective, culture is both the context *within* which and that *into* which the child develops (Lee & Walsh, 2001). Thus, as a systems perspective that focuses on "the role of contextual embeddedness and temporality in shaping the developmental trajectories of diverse individuals and groups" (Lerner, 2006, p. 6), cultural psychology takes seriously the idea that children develop within and into a specific cultural and historical context. Shweder et al. (1998) further explained:

> The wager of cultural psychology is that relatively few components of human mental equipment are so inherent, hard wired, or fundamental that their developmental pathway is fixed in advance and cannot be transformed or altered through cultural participation. The bet is that much of human mental functioning is an emergent property that results from symbolically mediated experiences with the behavioral practices and historically accumulated ideas and understandings (meanings) of particular cultural communities. (P. 867)

In summary, cultural psychology as a systems model emphasizes understanding children's development in context by paying careful attention to practices and their underlying values and meanings shared in a cultural community at a given historical time.

Examining the Discourse on ADHD from a Cultural Psychological Perspective

Although contemporary developmentalists, including cultural psychologists, emphasize paying attention to the context within which development occurs, the widely used diagnostic criteria for ADHD presented in the DSM (APA, 2000) do not seem to take the issue of context into serious consideration. As shown in table 2.1, the criteria simply stated that some symptoms and impairment should be present "in two or more settings" (p. 92). A careful examination of the manual reveals a few more remarks related to the context and that justify why observing symptoms in several settings is critical:

> Symptoms typically worsen in situations that require sustained attention or mental effort or that lack intrinsic appeal or novelty (e.g., listening to classroom teachers, doing class assignments, listening to or reading lengthy materials, or working on monotonous, repetitive tasks). Signs of the disorder may be minimal or absent when the person is receiving frequent rewards for appropriate behavior, is under close supervision, is in a novel setting, is engaged in especially interesting activities, or is in a one-to-one situation (e.g., the clinician's office). The symptoms are more likely to occur in group situations (e.g., in playgroups, classrooms, or work environments). The clinician should therefore gather information from multiple sources (e.g., parents, teachers) and inquire about the individual's behavior in a variety of situations within each setting (e.g., doing homework, having meals). (Pp. 86–87)

Although the intent of these descriptions might be to caution that one should not gather data about a child's behavior only in the situations that either aggravate or minimize features symptomatic of ADHD, the guidelines did not explicitly discuss how different expectations and meanings people in different settings have about children's behavior might influence ADHD diagnoses—one of the key issues that teachers often have to negotiate with children's parents. In relation to culture, the DSM very briefly mentioned: "[ADHD] is known to occur in various cultures, with variations in reported prevalence among Western countries probably arising more from different diagnostic practices than form differences in clinical presentation" (APA, 2000, p. 89). The manual, however, did not explain what causes different diagnostic practices in different cultural communities.

Not providing further discussion about these critical contextual issues, the manual went on to state that data "obtained from multiple informants (e.g., baby-sitters, grandparents, or parents of playmates) are helpful in providing a confluence of observations concerning the child's inattention, hyperactivity, and capacity for *developmentally appropriate* self-regulation in various settings" (APA, 2000, p. 91; emphasis added). This statement indicates that the DSM continued to frame ADHD as an abnormality identifiable through measuring children's behavior by the developmental norms.

These diagnostic criteria of ADHD clearly reflect the paradigm of the twentieth century's grand developmental theories in that they focus on identifying a disorder based on the universal, structural, individualistic, and dualistic notions of development. It is not surprising then that the early childhood teachers in my study tried to make sense of the diagnostic criteria and practices for ADHD based on these limited conceptions of development, which were familiar to them and in many ways reinforced by the authority on child psychiatry, the APA. In the remainder of this section, I examine the aforementioned limitations in the discourse on ADHD from a cultural psychological perspective.

First, both the DSM and the teachers' discussion about ADHD reflect universalism and structuralism through their assumption about developmental stages that children across time and place follow. The DSM assumed that everyone involved in ADHD diagnosis knows the norms of behavior at each age and developmental level. The teachers interpreted the norms as age-related behavioral features identifiable through comparison among the same age children and as age-graded skills measurable by standardized tests. In all cases, age was considered a natural marker for the process of biological maturation, and the norms were perceived as reflecting this universal process. In this discourse, neither the DSM nor the teachers framed using age to mark developmental stages and creating the norms around age as a cultural practice.

Rogoff (2003), however, found that in many cultural communities age is not even tracked. She argued that using age as a developmental marker is a recent practice even in the United States: "Before the end of the 1800s, people often did not know or have records of their birth date. It was not until the 20th century that Americans commonly referred to ages and began to celebrate birthdays regularly" (p. 155). This practice of using age for a developmental marker and of creating norms around age is an excellent example of what Bruner (1986) described as the process of reifying developmental theories in a culture:

> Theories of human development, once accepted into the prevailing culture, no longer operate simply as descriptions of human nature and its growth. By their nature, as accepted cultural representations, they, rather, give a social reality to the process they seek to explicate and, to a degree, to the "facts" that they adduce in their support. It is much as a theory of property is constitutive of concepts like ownership, trespass, and inheritance. By so endowing them with *social* reality, we give them a practical embodiment as well. (P. 134; emphasis in original)

Adopting Bruner's (1986) idea, we can see, once accepted as a habitual practice into the culture, the use of age for a developmental marker creates various social realities, such as various industries related to birthdays, birthday norms, laws on the driving age, and so on. In this sense, tracking children's ages, creating norms around ages, and measuring children's behavior by age norms become no longer universal and natural practices, but very much cultural ones. Likewise, from a cultural psychological perspective, the practice of ADHD diagnosis based on age norms is a cultural practice. Once accepted into the culture, ADHD creates social realities, such as various diagnostic procedures and tools, clinics, support groups (e.g., Children and Adults with ADHD), pharmaceutical companies, and so on.

Second, the discourse on ADHD focusing on biological and psychological problems reflects the individualistic and dichotomous notions of development by separating the individual from the context and by splitting biology and psychology from culture. For example, although the cause of ADHD is still unknown (NIH consensus statement, 1998), some scholars (e.g., Barkley, 2003) argue that ADHD results from a biological defect, such as a low level of dopamine in the brain (Singh, 2002a). Describing the potential heredity of ADHD, the DSM also stated, "Considerable evidence attests to the strong influence of genetic factors" (APA, 2000, p. 90). Singh explained that the biological approach to ADHD is the key to understanding the justification of medication treatment. Singh shared her correspondence with an expert who wrote: "the ideology of ADHD behavior as a brain disorder is so strongly entrenched in the U.S., that any study that might deny or delay the use of medication in the above age 6 years age group might be seen as medically unethical" (p. 579).

While the individualistic notion of development is widely spread by experts who use biological and medical discourse, this perception was also prevalent when the teachers framed the disorder as a psychological problem: "It seems like they want to be able to sit in their chair and listen, but they're just not able to... They want to really be able to, but they can't" (Kim, a third grade teacher). As described in this transcript, the teachers emphasized how ADHD children are unable to control their behavior at will.

These perspectives of ADHD focusing on biological, medical, and psychological issues disguise how the practice of diagnosis and medication treatment of ADHD as a cultural practice is linked to particular cultural values, expectations, and priorities. Rogoff (2003) argued that how soon children have to pass developmental "milestones" (e.g., walking, talking, reading) and what they are expected to do at each milestone reflect the value of cultural communities in which the children grow. For example, she described that, whereas European-American middle-class families emphasize children's early "verbal articulateness and assertiveness,...Italian signs of maturity focus on sensitivity to the needs of others and graciousness in entering and exiting social situations" (p. 159). Similarly, Bruner (1986) also pointed out,

> [The] truths of the theories of development are relative to the cultural contexts in which they are applied. But that relativity is not, as in physics, a question of logical consistency alone. Here it is also a question of congruence with values that prevail in the culture. It is this congruence that gives developmental theories—proposed initially as mere descriptions—a moral face once they have become embodied in the broader culture. (P. 135)

As these scholars articulated, from a cultural psychological perspective, the developmental norms are not the simple descriptions of the universal process of biological maturation, but the socially constructed framework or meaning system that reflects particular cultural values and expectations. In this sense, what is considered atypical development and a disorder needs to be examined in relation to cultural values.

As shown in table 2.1, the DSM defined ADHD as a disorder that should be treated because people with ADHD have "significant impairment in social, academic, or occupational functioning" (APA, 2000, p. 93). The teachers in my study worried that children with ADHD had academic malfunctioning that could negatively affect these children's future lives. Cautiously, all agreed that these children should get medication treatment to ensure their success in school. Success in school, from these teachers' perspectives, was defined as passing standardized tests measuring age-graded academic skills and as being ready for promotion to the next grade.

The participant teachers' concerns about standardized tests and readiness for grade promotion have cultural and historical resonance. Rogoff (2003) explained that age-graded schooling resulted from industrialization, which brought emphases on "efficiency and productivity...[and] order and predictability [in] human life" (Chudacoff, 1989, p. 5 as cited in Rogoff, 2003, p. 156). She described the history of age-graded norms and schooling:

> By the 1890s, concern with scheduling extended beyond the hours of the day to the years of life, as experts delineated norms for the ideal age timing of life events, prescribing what it meant for an individual's

experiences and achievements to be on time (or ahead or behind). Within a few decades, in the early 1900s, the interest in prescribing norms for the age of achievement of particular developmental milestones extended to concerns about characterizing individuals in terms of their degree of "retardation" (or "backwardness") versus "normal" development.

When schooling became compulsory, a standard starting age was required to enforce the schooling laws and catch truants. This allowed schools to move students through the grades in age "batches" given the same instruction. (P. 161)

Yet, the consideration of this culturally and historically constructed nature of norms and age-graded schooling and of their underlying cultural values is completely absent in the discourse on ADHD generated by the professional organizations, such as the APA and the NIH. As mentioned earlier, the DSM does not take different values and expectations people in different communities and at different historical times have about children's behavior into consideration for ADHD diagnoses. The teachers in my study acknowledged how today's school under *NCLB* has "become such a test-oriented school." However, this awareness did not lead the teachers to examine critically how this context might affect their views of children and ADHD.

Concerning the lack of recognition of the school as a culturally and historically constructed institution in psychological diagnoses, McDermott and Varenne (1996) argued,

> Cultures are not only occasions for disabilities, they actively organize ways for persons to be disabled... The problems that exist in one culture do not have to exist in another culture, or at least not with the same interpretations and consequences, and the same is true in the same culture at different points of its history or at different levels of its hierarchy. (Pp. 109–110; emphasis in original)

These researchers encourage educators to ask: Why is the same or similar behavior considered normal in one context, but abnormal in another? Why is the same or similar behavior, which was once perceived functioning, treated as a disorder in a different historical time within a culture?

In their study on children with a learning disability (LD), McDermott and Varenne (1999) analyzed these children's difficulties in connection to the underlying mechanism of American schooling that is run by competition and comparison among age-graded peers and that legitimizes the practice of psychological diagnosis as an ethical action.

Focusing on this cultural mechanism of American schooling, these researchers defined LD as "a cultural problem" (p. 27).

Varenne and McDermott (1999) argued that the individualistic notion of development reflected in the biological, medical, and psychological discourse on disabilities stems from the ideology of European American individualism. Echoing this idea, I argue that framing a disability as an individual trait (e.g., dopamine deficit, being unable to control one's own behavior) reflects the middle-class European American cultural belief about "the independent self" (Shweder et al., 1998), which characterizes the self as being "attribute-based (e.g., based in traits, preferences, goals)" (p. 901). Shweder and his colleagues elaborated that the study of self in Western social science has been rooted in "the ontology of individualism . . . The Latin word 'individual' means indivisible and whole, and the central tent of individualism is the epistemological priority accorded to the separate, essentially nonsocial individual" (p. 898).

Similarly to cultural psychologists' concern about understanding development in context, Varenne and McDermott (1999) urged educators to move beyond the individualistic view of development as stated in the following extract:

> [It] is our responsibility to move further aside to see more clearly the School as historically constructed, *always already there,* and always consequential. Above all, we must struggle against the ideological underpinning of this system when it tells us to center our gaze on the person as self-constituted individual. As one focuses on the learner, the focusing mechanism—America—disappears. Worse, the "individual" that appears alone, standing in isolation, thereby overwhelms the landscape and is yet subverted. The more attention paid to the individual, the more "determined" and the more restricted the person. To respect the individual, politically and morally, one must analytically cast one's eyes away. (P. 155; emphasis in the original)

Conclusions

In this chapter, I drew on data from my study of early childhood teachers' perspectives of children with ADHD (Lee, 2008) to illustrate how a limited developmental perspective based on the notions of developmental stages, norms, and readiness is reinforced by professional organizations, such as the APA, and is still dominant in the teachers' discussion about development and disability. I introduced cultural psychology (e.g., Bruner, 1996; Cole, 1996; Kitayama &

Cohen, 2007; Shweder et al., 1998) as a contemporary developmental systems model (Lerner, 1998, 2006) and discussed how this alternative framework might help early childhood teachers move beyond the limitations of the universal, structural, individualistic, and dualistic notions of child development stemming from the twentieth century's grand developmental theories.

I would like to note that my intent in this chapter is not to ask early childhood educators to maintain psychology as the sole knowledge basis by replacing developmental psychology with cultural psychology. Yet, I argue that, as much as the field's overreliance on one discipline is limiting, its exclusion of up-to-date perspectives from a particular discipline does not reflect our vision for reconceptualizing the field articulated by Bloch (1992): "Knowledge can come from many sources, and alternative ways of knowing can only add to our vision of issues, influences on development and schooling, and understanding of curriculum and pedagogy" (p. 16). I suggest cultural psychology as a contemporary developmental perspective be one of the many frameworks useful for our understanding of children's lives.

I believe cultural psychology has the great potential to be aligned with many ideas proposed by critical theorists. Also, the cultural psychologist's focus on the bidirectional relation between the individual and culture and on the plasticity of human development helps us recognize the "fundamental strength of each person" (Lerner, 2006, p. 11). Again, a human being "is not free of *either* his genome *or* his culture" (Bruner, 1986, p. 135; emphasis in the original), and his life is neither determined by his gene nor by his culture. I expect that cultural psychology's goal for appreciating "psychological pluralism" (Shweder et al., 1998) will contribute to leveling the playing field by facilitating more discussions and collaborative inquiry among early childhood educators in many cultures and in the many subcultures making up those cultures than providing prescriptions for the best practice based on one dominant cultural group's values and beliefs.

NOTES

Some of the discussions in this chapter are adapted from Lee & Walsh (2001) and Lee (2008).

1. Lee, K. (2001). *Raising the independent self: Folk psychology and folk pedagogy in American early schooling.* Unpublished doctoral dissertation, University of Illinois, Urbana-Champaign.

2. Dr. Daryl Scurry, a pediatrician, told me that most parents visit a physician to discuss their child's behavioral issues at the suggestion of their child's teacher (personal communication, March 18, 2006).

REFERENCES

American Psychiatric Association. (2000). *Diagnostic and statistical manual of mental disorders* (4th ed., text revision). Washington, DC: Author.

Barkley, R. A. (2003). Attention deficit hyperactivity disorder. In E. J. Mash & R. A. Barkley (eds), *Child psychopathology* (2nd ed., pp. 75–143). New York: Guilford Press.

Bloch, M. N. (1992). Critical perspectives on the historical relationship between child development and early childhood education research. In S. A. Kessler & B. B. Swadener (eds), *Reconceptualizing the early childhood curriculum: Beginning the dialogue* (pp. 3–20). New York: Teachers College Press.

Bredekamp, S. (1987). *Developmentally appropriate practice in early childhood programs serving children from birth through age 8*. Washington, DC: National Association for the Education of Young Children.

Bredekamp, S., & Copple, S. (1997). *Developmentally appropriate practice in early childhood programs* (revised ed.). Washington, DC: National Association for the Education of Young Children.

Bruner, J. (1986). *Actual minds, possible worlds*. Cambridge, MA: Harvard University Press.

———. (1996). *The culture of education*. Cambridge, MA: Harvard University Press.

Cannella, G. S. (1997). *Deconstructing early childhood education: Social justice and revolution*. New York: Peter Lang.

Cole, M. (1996). *Cultural psychology: A once and future discipline*. Cambridge, MA: Harvard University Press.

Copple, S., & Bredekamp, S. (2009). *Developmentally appropriate practice in early childhood programs serving children from birth through age 8* (3rd ed.). Washington, DC: National Association for the Education of Young Children.

Damon, W. (1998). Preface to the handbook of child psychology, fifth edition. In W. Damon (series ed.) & R. M. Lerner (vol. ed.), *Handbook of child psychology: Vol. 1. Theoretical models of human development* (5th ed., pp. xi–xvii). New York: John Wiley & Sons.

Einarsdóttir, J., Holst, J., Hong, Y., Kadota, R., & Lee, K. (April 2005). *Children with ADHD in five countries: Development and disability in culture*. Symposium at the annual meeting of the American Educational Research Association, Montréal, Canada.

Gibbs, N. (November 30, 1998). The age of Ritalin. *Time, 152*, 86–96.

Graue, M. E. (1993). *Ready for what? Constructing meanings of readiness for kindergarten*. Albany, NY: State University of New York Press.

Greenfield, P. (1997). Culture as process: Empirical methodology for cultural psychology. In W. Berry, Y. H. Poortinga, & J. Pandey (eds), *Handbook of cross-cultural psychology: Vol. 1. Theory and method* (2nd ed., pp. 301–346). Boston: Allyn & Bacon.

Holst, J. (2008). Danish teachers' conception of challenging behavior and DAMP/ADHD. *Early Child Development and Care, 178,* 363–374.

Hong, Y. (2008). Teachers' perceptions of young children with ADHD in Korea. *Early Child Development and Care, 178,* 399–414.

Jipson, J. (1991). Developmentally appropriate practice: Culture, curriculum, connections. *Early Education and Development, 2,* 120–136.

Kessler, S. A., & Swadener, B. B. (Eds). (1992). *Reconceptualizing the early childhood curriculum: Beginning the dialogue.* New York: Teachers College Press.

Kitayama, S., & Cohen, D. (Eds). (2007). *Handbook of cultural psychology.* New York: The Guildford Press.

Lee, K. (2008). ADHD in American early schooling: From a cultural psychological perspective. *Early Child Development and Care, 178,* 415–439.

Lee, K., & Neuharth-Pritchett, S. (2008). Attention deficit/hyperactivity disorder across cultures: Development and disability in contexts. Special Issue. *Early Child Development and Care, 178,* 339–346.

Lee, K., & Walsh, D. J. (2001). Extending developmentalism: A cultural psychology and early childhood education. *International Journal of Early Childhood Education, 7,* 71–91.

Lerner, R. M. (1998). Theories of human development: Contemporary perspectives. In Damon & Lerner (eds), *Handbook of child psychology: Vol. 1* (5th ed., pp. 1–24).

———. (2006). Developmental science, developmental systems, and contemporary theories of human development. In Damon & Lerner (eds), *Handbook of child psychology: Vol. 1* (6th ed., pp. 1–16).

Mallory, B. L., & New, R. S. (Eds). (1994). *Diversity & developmentally appropriate practices: Challenges for early childhood education.* New York: Teachers College Press.

McDermott, R. P., & Varenne, H. (1996). Culture, development, disability. In R. Jessor, A. Colby, & R. A. Shweder (eds), *Ethnography and human development: Context and meaning in social inquiry* (pp. 101–126). Chicago: The University of Chicago Press.

———. (1999). Adam, Adam, Adam, and Adam: The cultural construction of a learning disability. In H. Varernne & R. P. McDermott (eds), *Successful failure: The school America builds* (pp. 25–44). Boulder, CD: Westview.

Miller, P. (1998, February). *The cultural psychology of development: One mind, many mentalities.* Paper presented at the Social Development Consortium, University of Illinois at Urbana-Champaign.

NIH Consensus Statement. (1998). *Diagnosis and treatment of attention deficit hyperactivity disorder (ADHD).* Washington, DC: National Institutes of Health.

Nisbett, R. E. (2007). A psychological perspective: Cultural psychology—past, present, and future. In S. Kitayama & D. Cohen (eds), *Handbook of cultural psychology* (pp. 837–844). New York: The Guildford Press.

Rogoff, B. (2003). *The cultural nature of human development.* New York: Oxford University.

Ryan, S. K., & Grieshaber, S. (2005). Shifting from developmental to postmodern practices in early childhood teacher education. *Journal of Teacher Education,* 56, 34–45.

Shweder, R. A., Goodnow, J., Hatano, G., LeVine, R. A., Markus, H., & Miller, P. (1998). The cultural psychology of development: One mind, many mentalities. In Damon & Lerner (eds), *Handbook of child psychology: Vol. 1* (5th ed., pp. 865–937).

Singh, I. (2002a). Bad boys, good mothers, and the "miracle" of Ritalin. *Science in context,* 15(4), 577–603.

———. (2002b). Biology in context: Social and cultural perspectives on ADHD. *Children & Society,* 16, 360–367.

Spodek, B. (1988). Conceptualizing today's kindergarten curriculum. *The Elementary School Journal,* 89, 203–211.

Varenne, H., & McDermott, R. (1999). *Successful failure: The school America builds.* Boulder, CD: Westview.

Walsh, D. J. (1991). Extending the discourse on developmental appropriateness: A developmental perspective. *Early Education and Development,* 2, 109–119.

CHAPTER 3

Being Present in the Middle School Years

Hilary G. Conklin

"Dwelling in the present moment." While I sit here, I don't think of somewhere else, of the future or the past. I sit here, and I know where I am.

—Hanh, 1987a

An education professor once told me about a conversation she had with a group of middle school students. The professor was visiting the students' school and talking with them about what they enjoyed and disliked about their school. She described her discussion with these students:

> These were kids who pretty much like their school, but one of the things they said they didn't like is, they said, "well, why are the teachers constantly trying to scare us?" And I said, "well, what do you mean?" And they said, "well, you know everything you do, [the teachers say], well, they're not going to let you do this when you get to high school. And if you don't do this, you're not going to make it to high school." And this one young lady said it brilliantly—she said to me, "Why are they so focused on what's *going* to happen? Why not deal with what's happening? We're in school right now."[1]

As this student astutely observed, many middle grades (six–eight)[2] teachers are focused on preparing students for the future, for the education they are *going* to get—often to the exclusion of interesting, challenging learning they could engage their young adolescent students (ages ten–fourteen) in right now.

The phenomenon that this middle school student recounts in the extract given earlier is one that appears to be quite pervasive in American schooling. The education of young people in American society is permeated by an orientation on the future that often eclipses the value of learning in the present. As some authors have noted (Hanh, 1987a; hooks, 2003), we always seem to be waiting for something better to come. In this chapter, I explore how this orientation on the future often leads educators to postpone engaging, intellectually challenging learning until later when learners are "ready." I focus in particular on the ways in which teachers have made these arguments in relation to learners in the middle grades, the stage at which young people are "getting ready" for high school. Based on my own and other researchers' work, I argue that the future orientation that seems to pervade the middle grades—as well as other grade levels—may prevent students from ever having opportunities to take part in challenging, engaging learning.

Drawing on this concern, I suggest that Buddhist monk Thich Nhat Hanh's notion (1987a, b, 1993) of living in the present moment might provide a useful way for educators to rethink the conversations around the kind of learning that young adolescents should engage in during the middle school years. Hanh's guidance suggests that educators might look at the learners in their classrooms for who they are and where they are now—rather than who they might become and where they might go. By "dwelling in the present moment" (Hanh, 1987a, p. 6) with young adolescents, educators can come to see that young adolescents—just like all learners—are indeed ready for intellectually challenging learning.

I begin by providing a brief discussion of some of Thich Nhat Hanh's Buddhist perspectives and go on to orient the reader to my notion of "intellectually challenging" learning. I then provide several examples drawn from my own and other scholars' research to illustrate how educators have often argued for postponing more interesting learning until later, because learners are not yet ready. Drawing on these examples, I discuss how educators might use Hanh's orientation in order to reframe their way of thinking about the middle school years and, in doing so, provide learning experiences for young adolescents that are more engaging and challenging than what they often experience.

An Orientation on the Present

We tend to be alive in the future, not now. We say, "Wait until I finish school and get my Ph.D. degree, and then I will be really alive." When we have it, and it's not easy to get, we say to our-selves, "I have to wait until I have a job in order to be really alive." And then after the job, a car. After the car, a house. We are not capable of being alive in the present moment. We tend to postpone being alive to the future, the distant future, we don't know when. Now is not the moment to be alive. We may never be alive at all in our entire life. Therefore, the technique... is to *be* in the present moment, to be aware that we are here and now, and the only moment to be alive is the present moment. (Hanh, 1987a, P. 16; emphasis added)

In this quote, Hanh (1987a) speaks to the preoccupation of many industrialized societies with planning for the future and preparing for what might happen next, often to the exclusion of appreciating what is happening now. Indeed, this seems to be the very point that the young adolescent girl in the chapter's opening anecdote was making when she asked the questions, "Why are [the teachers] so focused on what's *going* to happen? Why not deal with what's hap-pening? We're in school right now." The American, age-graded schooling system—an artifact of industrialization (cf., Rogoff, 2003)—mirrors the societal focus on the future through its stepping stone pattern: every grade level represents preparation for something else, a constant ascension to something better. The numeric climb-ing of the grade levels suggests that seventh grade is of greater value than fifth grade, while ninth grade, in turn—the beginning of *high* school—is surely an even better place to be. This pattern, in turn, suggests to educators and students that the "good stuff" is always yet to come.

But Hanh's (1987a) guidance suggests that this focus on the future has critical limitations, because in the process of focusing on the future, as he puts it, "We may never be alive at all in our entire life" (P. 16). The alternative that he poses is that we shift to focusing on the opportunities we have right now. Hanh (1993) explains that "Only the present moment is real and available to us. The peace we desire is not in some distant future, but it is something we can realize in the present moment" (p. 5). Thus, instead of making educational decisions based on their value for what students will encounter at the next stage of education, Hanh's ideas suggest that educators bring their full attention to the students in each grade level, students who desire to learn, to be engaged, and to use their minds in the present moment. Rather than thinking of middle school as the time to pre-pare for high school, then, we can think of the opportunities for engaging learning that the middle school years afford.

One of the tools for being alive in the present moment is the practice of mindfulness. Hanh (1993) writes, "To be mindful means to be fully present in the moment" (p. 33). Being mindful means that if we are walking, we are aware that we are walking, rather than planning what we will do over the weekend. If we are eating, we pay attention to the food we are eating and the process of eating (Hanh, 1987b), rather than reading the newspaper or grading papers. This idea of mindfulness is part of engaged Buddhism and the principle of being in touch with oneself. As Hanh (1987a) explains, " 'in touch' means in touch with oneself in order to find out the source of wisdom, understanding, and compassion in each of us... to be aware of what is going on in your body, in your feelings, in your mind" (p. 85). Thus, it is through paying careful attention in the moment, to our own thoughts, feelings, and physical sensations that we can gain insight and greater clarity into ourselves, as well as other people and our surroundings. It is mindfulness that facilitates understanding. As Hanh (1987a) explains, "The root-word buddh means to wake up, to know, to understand..." (p. 13). In education, then, by paying attention in the moment, educators open up the opportunity to see their students more fully, and in doing so, can come to discover the true nature of the intellectually curious young people in front of them—young people who are eager and ready to learn.

INTELLECTUALLY CHALLENGING AND ENGAGING LEARNING

Jerome Bruner (1960) wrote that "any subject can be taught effectively in some intellectually honest form to any child at any stage of development" (p. 33). According to Bruner, learners of all ages can take part in substantive intellectual work, and they are all ready and able to do so, but it is the task of the educator to determine how to make that subject accessible to the student.[3] He explained his central conviction that

> intellectual activity anywhere is the same, whether at the frontier of knowledge or in a third-grade classroom. What a scientist does at his desk or in his laboratory, what a literary critic does in reading a poem, are of the same order as what anybody else does when he is engaged in like activities—if he is to achieve understanding. The difference is in degree, not in kind. (P. 14)

His critical point here is that, although the quantity or level of intellectual work may be adjusted to the particular learners, the nature of the intellectual work at different age or grade levels should be marked by comparable quality. Bruner suggests, then, that middle school students should engage in learning that is of the same intellectual quality as that of high school students, just as kindergartners should engage in learning of the same intellectual quality as middle schoolers, but this learning must be tailored to the particular learners present.

But what are the markers of intellectual quality? Newmann et al. (2007) provide some helpful defining features of what they term "authentic intellectual work," work that is characterized by rigorous, relevant learning that emphasizes higher order thinking skills and value beyond school. Learning with these characteristics demands that students—of any age—construct knowledge and manipulate information such that they arrive at new interpretations or conclusions. They are not simply receiving factual information or reciting prespecified ideas, but instead are explaining concepts and developing hypotheses. Students engaged in this kind of learning have opportunities to develop deep understandings of concepts and solve new problems, rather than learning fragmented, discrete pieces of knowledge. And, this kind of learning gives students the opportunity to talk with one another to develop shared understandings, rather than merely responding to teachers' questions. Finally, when students are engaged in relevant learning with intellectual quality, they have the opportunity to see connections between classroom learning and real world problems; they develop connections to their own lives and experiences.

Many researchers have documented that this kind of learning— what I refer to in this chapter as intellectually engaging or intellectually challenging—is indeed possible, as Bruner (1960) suggested, across the age levels. For example, in social studies education, scholars have studied classrooms in which elementary and middle school learners are interpreting historical evidence, drawing conclusions based on this evidence, or discussing controversial issues (e.g., Hess, 2002; VanSledright, 2002). Yet amidst this research evidence that documents the kind of learning that is possible with young learners, my own and other scholars' research suggests that many other educators are reluctant to have their students take part in this kind of learning because, in their view, students in the elementary or middle school grades are not ready. I now turn to this research and explore some specific examples in which Hanh's guidance (1987a, b, 1993) of focusing on the present might help educators think differently about the kinds of intellectual work students at all levels might engage in.

"NOT YET" IN THE ELEMENTARY YEARS

In the elementary grades, it seems that many educators are concerned that sophisticated intellectual work is either too difficult cognitively for young learners or is otherwise developmentally inappropriate. VanSledright (2002) made the observation that young children often have little opportunity to engage in sophisticated forms of historical investigation because teachers may misapply Piagetian notions of development. He notes:

> This creates an interesting self-fulfilling prophecy: We believe that children are incapable of difficult acts of historical thinking and investigation, so we prevent them from having opportunities to do so, which in turn reinforces our assumptions that they are incapable because we do not see them perform as such. (P. 8)

Indeed, in recent debates over the social studies curriculum in the state of Arizona, Hinde and Perry (2007) documented how K-3 educators used Piagetian theories and developmentally appropriate language to argue that the new curriculum was not appropriate for young children. As K-3 educators refuted the new standards, the concern they voiced most frequently was "that the standards were beyond primary-age children's stage of cognitive development" (p. 69). Hinde and Perry report that one first-grade teacher, for example, argued against the new standards by explaining that: "Challenging students to demonstrate comprehension of ancient, extinct civilizations halfway around the world is setting these students up for failure. No amount of teaching, resources, or interventions can cause a child to comprehend what s/he is not developmentally ready to comprehend" (p. 68). As VanSledright (2002) suggests, these early elementary educators did not want to even try to help young children learn particular social studies content because they feared students would not be capable of doing so.

James (2008) encountered similar arguments among preservice teachers in her elementary social studies methods courses as she tried to understand their resistance to teaching from multiple perspectives and using interpretation in the teaching of history. James explains:

> Consistently, as I engaged students in the practice and discussion of interpretation in history teaching, I encountered an interesting paradox: students seemed to agree that investigating history was more interesting and educational than the memorization-and-recall methods

they had encountered as students themselves, and yet students were disinclined to adopt such methods as teachers in their own classroom. (P. 172)

Her students told her that they felt the history teaching practices advocated in the course "would be both developmentally and morally inappropriate for elementary-aged children" (p. 173). These preservice teachers believed that "children cannot 'handle' critical thinking about historical content until they have ingested the 'basic facts' of a given time period" (p. 185).

While these early elementary teachers in James's study (2008) determined that the intellectually engaging practice of having students interpret historical evidence was not developmentally appropriate for young children, some of them did imagine that other teachers—those in the higher grades—might be able to try these teaching strategies. One of the preservice teachers in James's course, for example, explained, "If I were teaching upper elementary school this inquiry stuff would be more appropriate, but I hope to teach in the primary grades" (p. 187). Another preservice teacher in the course acknowledged that the class had read about a teacher who had successfully engaged fifth graders in historical interpretation (VanSledright, 2002), and thus, she concluded that "Maybe by [fifth grade] it would work, but not any earlier" (p. 187).

The studies I described above indicate that many elementary educators believe that certain kinds of intellectual work should be postponed until later grades—the upper elementary school years—when young children might be more ready for this kind of learning. In other words, these educators believe that challenging intellectual work should be reserved for some time later in the future—just not yet. But what happens when students arrive at that future? As I discuss later, it turns out that many educators repeat these same arguments in the upper grades: many secondary teachers believe that challenging learning with greater intellectual quality should be reserved for a new future—high school.

Middle School as a Stepping Stone to Real Learning

A few years ago, I conducted a case study at a research university in the upper Midwest to investigate how preservice teachers were prepared to become middle school social studies teachers depending upon whether they were prepared through an elementary or a secondary teacher education program (see Conklin, 2008, 2009). In this study, I tracked a cohort of elementary preservice teachers and another of secondary preservice teachers across one semester of coursework and field experience in their respective teacher education programs, using classroom observations, interviews, surveys, and document analysis to understand their learning about teaching social studies at the middle school level.

One of the striking findings of this research was that the secondary teachers, who were earning certification for teaching in grades six through twelve, were much more likely to hold low expectations for the intellectual work that young adolescents could accomplish than the elementary teachers, who were earning certification for teaching in grades three through eight. Although the secondary methods course had repeatedly emphasized pushing students to do challenging, intellectually rich social studies that requires higher-order thinking, some of the secondary teachers decided that this kind of work simply didn't apply to their middle school students. Many of the secondary preservice teachers espoused challenging social studies teaching strategies that would engage students in higher-order thinking and interesting learning—but believed that these strategies were more appropriate for high school students. Meanwhile, these secondary teachers felt that their middle school students should engage in less demanding work that would prepare them for the more rigorous (and interesting) work they could accomplish in high school.

The lower expectations that the secondary teachers held for young adolescents seemed to be part of their broader way of thinking and talking about the middle school years as a stepping stone to high school. Their language suggested that middle school is not a time in its own right, but rather a time to prepare for something else. For example, Brett,[4] one of the secondary preservice teachers explained her view of the middle school years, noting that "you're not just a little kid in Mr. Nelson's class anymore, but then again you're not like an AP kid or like a basketball kid." Her comment implies that students in elementary school and high school possess particular identities, yet during the middle school years, students enter an identity-less, no-person's-land.

For some of these secondary preservice teachers, the idea of the middle school years as a stepping stone was mirrored in the way they envisioned the middle school social studies curriculum, that the curriculum during the middle grades was a time to prepare for something else—something better. Will, for example, explained his ideas about the reasons for studying social studies in middle school:

> I think when you get to high school it's a little more about problem solving and becoming active in your community. Maybe in middle school it's a little bit more about learning about your community...not so much about getting totally involved in it right away. Just learning about it so you can find where you can get involved and, kind of build up to handle the problems that you're going to face. Not that I think that kids can't handle it at that age, but they need the background information first...It's kind of a base-building period, maybe.

Will perceived that middle school is not necessarily the time for students to take action in the context of social studies and their community, but it is a time to get ready for this. And, like one of the teachers in James's study (2008), Will indicates that middle school students should learn facts first—background information—before they can engage in actual problem solving. Similarly, Brett commented on what she saw as the reasons for studying social studies in middle school:

> I hate to cheat middle school social studies teachers, but I think it's basically an introduction to high school. I think it's introducing students about being American citizens, a lot of the time. About issues like voting and budgets and court system, and just the basics. Like, what is a Constitution? I think that's why you would have one course that's geography and one course that's U.S. History...they're just kind of introductions...So, I really think it's just to get them to know, first about their own area around them. I don't think you're going to get any impressive, intensive content in any [middle school] social studies course.

For both Brett and Will, then, the learning that goes on in middle grades social studies is not necessarily significant or interesting in its own right, but only because it serves to prepare students for the more interesting kinds of work they can engage in when they get to high school.

The future-oriented language of these two preservice teachers reflects that of their cooperating teachers. Will's cooperating teacher Ms. Webb explained that "We tell our kids that they're going to get all their subject matter again in high school, and our job in middle school is to get them ready for high school." Similarly, at a school on the other side of town serving a very different demographic of students, Mr. Briggs, Brett's cooperating teacher, commented:

> Middle school in general, we're kind of getting them ready for high
> school. That's kind of our goal. And especially 8th grade. With just
> the skills—if it's writing and reading skills or note-taking skills or even
> just behavior and responsibility and learning respect... It's just a mat-
> ter of us kind of rounding them out a little bit more and getting them
> prepared.

These appear to be common mantras that teachers hear and repeat:
middle school is preparation for high school.

This is also an idea that is reflective of and reinforced in large
part by the existing school curriculum. As VanSledright (2002)
points out:

> An examination of the typical survey American history course sequence
> and pedagogical approach that students experience repeatedly in U.S.
> schools (in grades 5 and 8, and at some point in high school) suggests
> that curriculum policy makers view late high school as the point where
> this deeper exploration is to occur, if at all (e.g., Cuban, 1991; Goodlad,
> 1984; Tyack & Tobin, 1994). According to this view, it takes until
> then to build a large enough store of historical knowledge to do more
> than reproduce it on recall tests. (P. 8)

Thus, many teachers' vision of what is appropriate for middle school
students may be shaped in part by the nature of the existing
curriculum.

Further, just as elementary teachers in different contexts have
made similar arguments about the kinds of learning young students
should and shouldn't be engaging in, the findings I described are
not unique to the particular secondary teachers in the examples. For
instance, in a more recent study situated in a research university in
the Deep South, my colleagues and I found the same phenomenon
among beginning secondary social studies teachers (Conklin et al.,
in press). In this project, we studied the influence of a series of structured course interventions on beginning secondary teachers' views
of young adolescents' intellectual capabilities. Prior to the course
interventions, we surveyed and interviewed the beginning teachers
in order to find out their initial conceptions of young adolescents'
capabilities. Much as I had found before, many of the beginning
teachers initially expressed skepticism that young adolescents could
do much more than engage in learning factual knowledge. These
teachers suggested that higher order thinking should be reserved

primarily for the high school years. For example, in a focus group interview prior to the structured coursework, one beginning teacher Isaac explained:

> I think a lot of higher level social studies gets into the abstraction of…certain concepts like citizenship…and supposedly, psychologically, I didn't think those skills were developed until a little bit after middle school age, so I just wouldn't think that the high level stuff would be as attainable…

Another participant Leah expressed a similar perspective as she explained her response to a survey question that asked about the best approach to teaching seventh grade social studies (see table 3.1). Leah explained:

> As far as controversial public issues and examining stuff like that…engaging minds. I think that usually works better when you are older and have been able to really live life a little more to formulate your own opinions, and I just think that maybe the goal of history in middle school is to prepare them for higher level work by giving them basic facts and helping them to start learning to build their own opinions.

These comments echo the notion that young adolescents need to do the preparatory work in middle school that will allow them to do more interesting intellectual work when they get to high school.

Table 3.1 Teachers' approaches to teaching seventh grade social studies

Three teachers, Pat, Lou, and Chris, describe their approach to teaching seventh grade social studies. Which of these teachers do you think has the best understanding of how to teach social studies to middle school students?

Pat: "In my class, students discuss controversial public issues, examine primary sources, and analyze social problems. These are the things that engage middle school minds."

Lou: "In my class, I work hard at helping students learn some of the fundamental ideas and events from our history. I want to give them the critical background knowledge that will enable them to do high level work when they get to high school."

Chris: "In my class, I want to make social studies come alive by getting students involved in hands-on activities. For example, I have my students make food from different cultures and build models of historical structures like the Roman Coliseum."

The Hormonal Challenge

The research I described above is not unique in finding that beginning secondary teachers often view the middle school years as a time that is not optimal for engaging students in intellectually challenging learning. Additional research indicates that, in particular, many secondary teachers believe that young adolescents' developmental level and active hormones preclude them from taking part in interesting intellectual work. For example, in another study of secondary preservice social studies teachers, Yeager and Wilson (1997) examined how a social studies methods course could help student teachers learn to use historical thinking tools that involved higher-order thinking in their secondary social studies teaching. These researchers found that the student teachers in their study used students' maturity level as a rationale for not using the historical thinking teaching strategies they had learned. The researchers explained:

> Another factor that appeared to be significant for a few student teachers was the maturity level of their pupils. Their concerns seemed to stem from particular classroom-management problems with early adolescents and their doubts about these pupils' intellectual capabilities...student teachers at middle schools were less likely than their high-school counterparts to engage in historical-thinking exercises with their pupils. The middle-school student teachers apparently believed that developmental and behavioral concerns hindered their creativity in designing more pupil-centered, inquiry-based lessons. (P. 125)

Similarly, Lexmond (2003) conducted a case study of secondary teachers being prepared for middle and high school teaching in a general methods course who took part in four curricular activities focused around eliciting and then challenging their assumptions about young adolescents and middle school teaching. Lexmond drew upon the work of Finders (1999) who had found that preservice teachers have views of young adolescents as "mindless entities, as detached hormones tied to uncontrollable bodies" (p. 256). In her study, Lexmond found that preservice teachers thought it was the exception rather than the rule to have young adolescents who were interested and intellectually capable: the preservice teachers thought that young adolescents with intellectual interests were outside of the norm. Following her curricular interventions, Lexmond found that although "some of the characterizations of early adolescents were 'disturbed' for the key pre-service teachers during the course of the semester, this notion of middle school students as intellectually incapable was not" (p. 47). According to Lexmond's research, these teachers continued to view young adolescents primarily in terms of their biological development, a perspective that led them to negative generalizations about the intellectual potential of students in the middle grades.

While the studies I described above provide research evidence to illustrate many secondary teachers' entrenched views of the kind of intellectual work that is possible in the middle school years, the casual observer need look no further than the portrayal of middle school students in popular culture to find evidence for the idea that, as Lesko (2005a) observes, "adolescents are considered under the control of hormones and unavailable for serious (i.e. critical) school tasks and responsibilities" (p. 98). Take, for example, a series of articles by the *New York Times* on "The Critical [Middle School] Years." The titles of the articles in this series include "Middle school manages distractions of adolescence" (Hu, May 12, 2007); "For teachers, middle school is test of wills" (Gootman, March 17, 2007); and "Trying to find solutions in chaotic middle schools" (Gootman, January 3, 2007). In reading these headlines alone, it would be fair to ask: What teacher would want to attempt meaningful intellectual work amidst these distractions, this chaos, and the test of wills?

Further, in a subsequent news story, Gootman (December 25, 2008) wrote that "throughout the country, middle school is increasingly seen as a kind of educational black hole where raging hormones, changes in how youngsters learn and a dearth of great teachers can collide to send test scores plummeting" (p. 5). It appears to be too hard to resist reinforcing the stereotypes of raging hormones and intellectual voids in early adolescents. Gootman goes on to say, "Many parents fear that picking the wrong school could dash their children's chances for a top high school or college" (p. 6). (Because, remember, middle school is preparation for high school.) With this journalistic drama being broadcast to millions of readers daily, it is no wonder then that beginning secondary teachers come to think of the middle school years as a time when it might be difficult to engage students in intellectual pursuits.

Looking across these different contexts of elementary and secondary teaching, and looking at the arguments made about learners in both research and popular culture, we should be deeply concerned for the education of young people. Elementary teachers think intellectually challenging learning should be postponed until the upper elementary years, yet secondary teachers think this kind of learning is more appropriate for high school. In both cases, teachers argue that students need to learn the facts first, build their background knowledge, and then do some interesting intellectual work later, once learners are more ready. But if all of these educators across the grade levels are making these same arguments, will students ever get to participate in learning that is engaging, thought-provoking, and intellectually challenging? One only needs to examine the crisis in high school dropout rates to realize that, for many students, the answer is no (Bridgeland, DiIulio, & Morison, 2006).

REORIENTING ON THE PRESENT

Given this well-documented problem across contexts that educators often believe students are not ready for learning that engages them in higher-order thinking and problem-solving, how might we rethink the idea of "getting ready" for more sophisticated, better learning? I now consider how Thich Nhat Hanh's (1987a, b, 1993) perspective might inform the work of teaching at all levels through his emphasis on living in the present moment.

To begin with, as educators, we might shift our thinking to focus on who young adolescents are right now, rather than who they might become at a later date or where they might be going. It seems, for example, that the view of young adolescents as not yet ready for challenging learning is reinforced by characterizations of the middle school years that emphasize the shifting nature of the adolescent terrain. The common description of the middle school years as a time of transition or "Turning Point" (cf., Jackson & Davis, 2000), while a well-warranted description in many ways and productive for highlighting the importance of these years of human life, unfortunately seems to have drawn attention away from the actual "point" around which the turning occurs. The focus of a "transition" is on movement, rather than on what is happening right now, in the present moment (Hanh, 1987a). By continuing to rely on the notion of young adolescence as a transition, we seem to rob young adolescents of who they are at particular moments in time.

If we shift our orientation to the present, when a beginning teacher like Brett explains that young adolescence is a time when "you're not just a little kid in Mr. Nelson's class anymore, but then again you're not like an AP kid or like a basketball kid," we might pose the question, but who *are* the young adolescents in front of us, rather than asking who or what they are *not*? We can ask, what kind of people are our students right now, rather than thinking in terms of who they have been or the people they are transitioning into? What are their interests, their passions, their concerns? What are their questions about the world? (Brodhagen, 1995)

If we can think of young adolescents in the present moment, as really alive (Hanh, 1987a) in our classrooms, then this means we need to seize all of our opportunities for learning. As Hanh (1993) reminds us, "Only the present moment is real and available to us" (p. 5). Brett explained, "I hate to cheat middle school social studies teachers, but I think it's basically an introduction to high school. I think it's introducing students about being American citizens..." If we are only introducing in the middle school years, or preparing young adolescents for the learning they might take part in at the high school level, then we are never getting to the good stuff—the stuff that actually engages students and challenges them to use their minds. Instead of getting them ready for high school, as so many educators seem to be inclined to do, we should focus on the exciting learning that is possible *today*, in *this* class period, in *this* moment. If we don't do this, as Hanh (1987a) warns, "We may never be alive at all in our entire life" (p. 6), and indeed, it is the lack of opportunities to be intellectually engaged and alive that many high school dropouts cite as a central reason for leaving school early (Bridgeland et al., 2006).

READY RIGHT NOW

Some skeptics may argue that focusing on who young adolescents are right now will only reveal that they are not ready to engage in challenging intellectual work. Yet it is here that Hanh's ideas (1987a) about mindfulness and paying full attention in the present moment can be helpful. Hanh explains that "Buddhism teaches us how to look at things deeply in order to understand their own true nature..." (p. 35). In fact, just as Ayers (2001) has written about the importance of observing children carefully in order to see them in terms of their strengths rather than from a deficit perspective, Hanh's ideas suggest that many educators may not be fully aware of young adolescents' true nature. He argues that we often have misperceptions about other people that can lead to human suffering. Hanh (1987a) explains:

> In Buddhism, knowledge is regarded as an obstacle to understanding, like a block of ice that obstructs water from flowing. It is said that if we take one thing to be the truth and cling to it, even if truth itself comes in person and knocks at our door, we won't open it. For things to reveal themselves to us, we need to be ready to abandon our views about them. (P. 42)

As educators, then, it may be easy to cling to the cultural myths about adolescence that highlight middle school learners' "physiological turmoil" and how these young people are "hormonally burdened" (Lesko, 2005a, p. 88), because these perceptions are reinforced across multiple contexts, in the language of practicing teachers and in the popular media. Yet these are the very views that we as educators need to be ready to abandon.

Paying careful attention to young adolescents reveals that these are students who have deep intellectual curiosity and an eagerness to engage these curiosities (cf., Beane, 1993; Brodhagen, 1995). As Beane (1993), for example, notes: "Those who really listen to early adolescents know that at both personal and social levels many are concerned about the environment, prejudice, injustice, poverty, hunger, war, politics, violence and the threat these issues pose to the future of our world" (p. 18). Looking at young adolescents through this new lens opens up new possibilities for learning with them. In my research, preservice teacher Will theorized that "Maybe in middle school it's a little bit more about learning about your community...not so much about getting totally involved in it right away. Just learning about it so you can find where you can get involved and, kind of build up to handle the problems that you're going to face." Yet young adolescents are already aware of problems in their communities and are facing problems of their own. They notice inequities inside school and out, and they are eager to learn about the problems that the world faces. They are concerned about global warming, about conflict around the world, about poverty. As Saltman (2005) notes:

> a critical approach encourages us to ask how it is, for example, that adolescents are expected not to know about political realities yet these same political realities keep roughly 40 million youth in the United States living in dire poverty with the average age of a homeless person in America standing at 9. (P. 19)

It is up to educators to decide whether to engage young adolescents' intellectual interests and concerns or not. If we do not engage them in solving significant problems and thinking through meaningful solutions, they will come up with their own solutions to avoid the mind-numbing school tasks with which they are faced or to accomplish these tasks with as little pain as possible.

Some readers might ask, however: Is it not true that young adolescents are going through a transition, that they are indeed in the midst of profound physiological, cognitive, emotional, and social changes?

If we are going to understand their "true nature," aren't hormones a part of this reality? Saltman (2005) provides a helpful explanation here about the socially constructed nature of adolescence. He explains that while there are particular physical realities that we can point to, the social construction of adolescence means that biological and psychological facts gain meaning through the ways these facts are interpreted in particular social contexts. He says:

> To claim that adolescence is a social construction is not to say, for example, that puberty is a fiction or merely a narrative with no natural scientific content. However, to recognize that adolescence is a social construction is to recognize... that the meaning of biological or psychological realities do become meaningful or relevant in different ways in different social contexts. (P. 16)

Thus, educators can use young adolescents' physiological attributes as their defining characteristics and think of them primarily in terms of their hormones and their affinity for peers (Lesko, 2005b), or they can think of young adolescents first as learners, people with ideas and questions about the world, who need educators who can connect their current understandings to challenging opportunities to solve problems. This may mean recognizing that young adolescents often struggle with organization, or possess a great deal of energy, or are interested in interacting with their peers. But as educators we have a choice of whether to cling to these physiological characteristics and the stereotypes these characteristics often accompany, or whether to acknowledge these facts as partial truths about young adolescents that can inform our instruction and meanwhile continue to pay attention to these young people so that we can more fully understand their true nature (Hanh, 1987a) and engage them in challenging learning.

CONCLUSIONS

I have argued throughout this chapter that Thich Nhat Hanh's ideas (1987a, b, 1993) about paying full attention to the present moment can be a productive way to rethink the common ideas in American education that intellectually challenging learning should be saved until the future. By focusing on the present, educators at all levels have responsibility for engaging learners in interesting intellectual work, because not seizing the present moment risks the possibility that learners will never have the opportunity to take part in work that challenges and engages them.

Before concluding, it is important to note that I do not want readers to interpret my use of Hanh's ideas here as a short-sighted method of "living in the moment" without regard to future consequences. As I hope has become clear by this point, this is not an argument to be careless or take a "nothing matters" approach to educating young adolescents. To be sure, we want middle school students to have engaging, interesting middle school years, as well as engaging, interesting high school years, higher education, and adulthood. It is also not to say that the learning a student engages in at one point in her education will not serve her at a later point in her education. And, lest the reader think that I have skillfully extricated myself from the grips of American society's forward-focused ideology, as I sit here writing this I am reminded of how I work in a field called "teacher preparation," and I must continually practice the very habits with the learners in my classroom that I advocate for other educators. I, too, have told my students that I want to make sure they are prepared for their future classrooms or their future careers. And I care about what happens to learners next.

But the point is, we need to care particularly about what happens to learners right now. We need to dwell in the present moment, to not focus on "the future or the past" (Hanh, 1987a, p. 6). And we can hope that in that moment we might be lucky enough to have students who will remind us, "We're in school right now."

NOTES

1. Conklin, H. G. (2006). *Learning to teach social studies at the middle level: A case study of preservice teachers in the elementary and secondary pathways.* Unpublished doctoral dissertation, University of Wisconsin-Madison School of Education.
2. In this chapter, I use the phrases "middle grades" and "middle school years" interchangeably to refer to grades six through eight.
3. Although the concern that Bruner speaks to here is that of educators *delaying* challenging learning until later, my colleague Kyunghwa Lee has also pointed out that some educators have used Bruner's ideas to justify teaching children skills *in advance* using traditional or behavioristic approaches. For example, many children and adolescents in Korea and other Asian countries attend "cram institutions" (after-school programs) where they learn academic skills in advance of the grade level they are about to enter, such as first graders learning about second grade subject matter one year prior to becoming second graders. Hanh's ideas, then, are useful for reminding educators to pay attention to the present moment in cases when educators are inclined to delay challenging learning until later as well as in cases when educators are inclined to

THE MIDDLE SCHOOL YEARS

teach academic skills in advance. In both situations, educators are focused on the future and losing sight of the present moment.

4. All participant names are pseudonyms.

REFERENCES

Ayers, W. (2001). *To teach: The journey of a teacher* (2nd ed.). New York: Teachers College Press.

Beane, J. A. (1993). *A middle school curriculum: From rhetoric to reality* (2nd ed.). Columbus, Ohio: National Middle School Association.

Bridgeland, J. M., DiIulio, J. J., Jr., & Morison, K. B. (2006). *The silent epidemic: Perspectives of high school dropouts.* A report by Civic Enterprises in association with Peter D. Hart Research Associates for the Bill & Melinda Gates Foundation.

Brodhagen, B. (1995). The situation made us special. In M. W. Apple & J. A. Beane (eds), *Democratic schools* (pp. 83–100). Alexandria, VA: Association for Supervision and Curriculum Development.

Bruner, J. S. (1960). *The process of education.* New York: Vintage Books.

Conklin, H. G. (2008). Promise and problems in two divergent pathways: Preparing social studies teachers for the middle school level. *Theory and Research in Social Education, 36*(1), 591–620.

———. (2009). Purposes, practices, and sites: A comparative case of two pathways into middle school teaching. *American Educational Research Journal, 46*(2), 463–500.

Conklin, H., Hawley, T., Powell, D., & Ritter, J. (in press). Learning from young adolescents: The use of structured teacher education coursework to help beginning teachers investigate middle school students' intellectual capabilities. *Journal of Teacher Education.*

Finders, M. J. (1999). Raging hormones: Stories of adolescence and implications for teacher preparation. *Journal of Adolescent & Adult Literacy, 42,* 252–263.

Gootman, E. (January 3, 2007). Trying to find solutions in chaotic middle schools. *New York Times.* Retrieved January 5, 2007, from http://www.nytimes.com/2007/01/03/education/03middle.html.

———. (March 17, 2007). For teachers, middle school is test of wills. *New York Times.* Retrieved March 18, 2007, from http://www.nytimes.com/2007/03/17/education/17middle.html.

———. (December 25, 2008). All's fair in the middle school scramble. *New York Times.* Retrieved December 25, 2008, from http://www.nytimes.com/2008/12/26/education/26fifth.html.

Hanh, T. N. (1987a). *Being peace.* Berkeley, CA: Parallax.

———. (1987b). *The miracle of mindfulness.* Boston: Beacon.

———. (1993). *Interbeing.* Berkeley, CA: Parallax.

Hess, D. E. (2002). Discussing controversial public issues in secondary social studies classrooms: Learning from skilled teachers. *Theory and Research in Social Education, 30*(1), 10–41.

Hinde, E. R., & Perry, N. (2007). Elementary teachers' application of Jean Piaget's theories of cognitive development during social studies curriculum debates in Arizona. *The Elementary School Journal, 108*(1), 63–79.

hooks, b. (2003). *Teaching community: A pedagogy of hope.* New York: Routledge.

Hu, W. (May 12, 2007). Middle school manages distractions of adolescence. *New York Times.* Retrieved May 18, 2007, from http://www.nytimes.com/2007/05/12/education/12middle.html.

Jackson, A., & Davis, G. A. (2000). *Turning points 2000: Educating adolescents in the 21st century.* New York: Teachers College Press.

James, J. H. (2008). Teachers as protectors: Making sense of preservice teachers' resistance to interpretation in elementary history teaching. *Theory and Research in Social Education, 36*(3), 172–205.

Lesko, N. (2005a). Denaturalizing adolescence: The politics of contemporary representations. In E. R. Brown & K. J. Saltman (eds), *The critical middle school reader* (pp. 87–102). New York: Routledge.

———. (2005b). Back to the future: Middle schools and the Turning Points report. In Brown & Saltman (eds), *The critical middle school reader* (pp. 187–195).

Lexmond, A. J. (2003). When puberty defines middle school students: Challenging secondary education majors' perceptions of middle school students, schools, and teaching. In P. G. Andrews & V. A. Anfara, Jr. (eds), *Leaders for a movement: Professional preparation and development of middle school teachers and administrators* (pp. 27–52). Greenwich, CT: Information Age.

Newmann, F. M., King, M. B., & Carmichael, D. L. (2007). *Authentic instruction and assessment: Common standards for rigor and relevance in teaching academic subjects.* A report prepared for the Iowa Department of Education.

Rogoff, B. (2003). *The cultural nature of human development.* New York: Oxford University Press.

Saltman, K. J. (2005). The social construction of adolescence. In Brown & Saltman (eds), *The critical middle school reader* (pp. 15–20).

VanSledright, B. (2002). *In search of America's past: Learning to read history in elementary school.* New York: Teachers College Press.

Yeager, E., & Wilson, E. (1997). Teaching historical thinking in the social studies methods course: A case study. *Social Studies, 88*(3), 121–127.

CHAPTER 4

Am I a Novice Teacher? The Voices of Induction Teachers in a Preschool

Su Kyoung Park and Amy Noelle Parks

In this chapter, we take up developmentalism and readiness in an educational context where they are rarely openly discussed: teacher education. Although research on developmentalism has not shaped the field of teacher education in the same ways as the construct has shaped the fields of early childhood and middle grades education, the notion that people become teachers by passing through a series of relatively predictable stages is quite common in the field (e.g., Costigan, 2004; Metzler et al., 2008). In addition, teacher educators routinely attempt to determine whether novices are "ready" for particular benchmarks, such as student teaching, certification, or independent teaching, in their own classrooms.

This chapter focuses on the experiences of three induction teachers in early childhood classrooms in South Korea. This context is illuminating for two reasons. First, the strong discourses of developmentalism and readiness in early childhood education reinforce similar ways of thinking about the learning and growth of novice teachers. Some of the earliest work on the stages of teacher development came from early childhood educators explicitly using research about the development of children as a way of thinking about the growth and learning of beginning teachers (e.g., Katz, 1972). Second, in Korea, even more than in the United States, there exists a strong cultural belief in the need to respect those who are older and more experienced. This cultural belief makes the consequences of developmental thinking about novice teachers more significant and easier to identify. By examining the ways in which developmental thinking about readiness impacts the experiences of novice teachers in Korea, we hope to provide an analytic framework that could be used to reinterpret the experiences of novice teachers in the United States, where hierarchies between novices and experts may be less openly recognized or discussed.

INDUCTION AS A STAGE

We write this chapter primarily in response to stage literature in teacher education and to theories of learning that position novices as outsiders or as peripheral participants. In recent years, much has been written about teacher induction (e.g., Hebert & Worthy, 2001; Howe, 2006; Stansbury & Zimmerman, 2000; Wayne et al., 2005). Much of this writing, even when it does not discuss developmentalism, portrays induction as a particular stage in development bridging preservice and expert teaching, much in the same way that adolescence bridges childhood and adulthood (e.g., Feiman-Nemser, 2001).

Much like early childhood and adolescence, induction is a constructed stage, which became possible to talk about and to study only relatively recently (Horn et al., 2002). The proliferation of literature on this subject has created a number of expectations about what it means to be an induction teacher and has worked to name beginning teachers as people with particular qualities and concerns that differ from those of teachers who have been in the classroom for longer. For example, early induction work has identified novice teachers as being preoccupied with management, parent relationships, and administrative tasks and has suggested that new teachers were unable to tackle more complex concerns related to topics such as curriculum and assessment until later in their careers (Bloom et al., 1991; Katz, 1972; Veenman, 1984). More recently, Sabar (2004) wrote that novice teachers must leave a familiar culture and move into a strange one where "the illusions, the hopes and expectations, the despair, the crises, the sense of loss and grief [are] replaced by compromise, acceptance and adjustment" (p. 146). What these arguments all have in common is the notion that induction (like adolescence) is an unavoidable and troubling stage that must be endured before one can emerge in a more advanced state. This stance can be seen in the emphasis on problems and struggles in the induction and mentoring literature (Berson & Breault, 2000; Brock & Grady, 1998; Lortie, 1975).

While beginning teachers do have very real concerns, we would like to suggest that the framing of induction as a unique stage implies that the troubles and concerns experienced by new teachers are quite different than those experienced by teachers who have worked for longer. The emphasis on problems and struggles makes it difficult to see the strengths that beginning teachers bring to their classrooms and schools and to analyze the ways that these newcomers might productively impact their colleagues and schools.

In addition to reframing beginning teachers as people who share concerns with more experienced teachers and who bring strengths into the profession, we would also like to open a discussion about theories of learning that portray novices as peripheral participants in communities they enter. In Lave and Wenger's work (1991; Wenger, 1998) on communities of practice, learning is described as increasingly central participation in a community, where experts exist at the center of the community and novices exist on the outer edges. Although some newcomers may consciously choose to remain peripheral participants, most beginners are seen as having an "inbound trajectory that is construed by everyone to include full participation in its future" (Wenger, 1998, p. 166). The time of nonparticipation during early entry to the community is seen as a necessary time of learning that will support full participation later on. This theory of learning has proved useful to a number of educational researchers who have looked at teacher learning in various settings, particularly in communities that emphasize collaboration (e.g., Flores, 2007; Morrell, 2003). Our concern with framing learning in this way is that it positions newcomers to the community as people who desire or are seeking to linearly move toward expert practice. We would like to open up the possibility that new teachers may chose a learning trajectory that positions them toward practices other than those that are valued by the experts in the community in which they are located and even to consider the possibility that newcomers may seek to alter or move the center of the community. In this way, the learning of new teachers might be seen as something other than enculturation into a community.

To consider teacher learning in this way, we draw on the work of literary critic Bakhtin (1993) who argued that each person has a unique place in Being. In writing about "Being" in the world, Bakhtin stated:

> The world is arranged around a concrete value-center, which is seen and loved and thought. What constitutes this center is the human being: everything in this world acquires significance, meaning, and value only in correlation with man—as that which is human. All possible Being and all possible meaning are arranged around the human being as the center and the sole value. (P. 61)

From this perspective, it is not others or experts who are central to our learning trajectories, but our own interpretations of our unique experiences with others in the social world. In everyday life, teachers experience countless "events" as they engage with others in the world around them. As human beings, teachers then assign meaning to these events. This assignment of meaning, which can vary from person to person, provides a way for thinking about learning and change where the trajectory for learning, even within the same community, might be quite different for different individuals. For Bakhtin, "a value-judgment about one and the same person that is identical in its content ('he is bad') may have different actual intonations, depending on the actual, concrete center of values in the given circumstances" (p. 63). Similarly, notions of what it means to be a good teacher, a learning teacher, or even a novice or experienced teacher, are dependent on continually evolving circumstances. Thus, learning can be understood as individuals' decisions to position themselves and to take action in light of the social ideologies and discourses in which they operate. A teacher could be seen to learn through her resistance to expert practice in addition to her movement toward it.

The Korean Context

As mentioned in the introduction, the categories of novice and expert are particularly important in the Korean context. Confucian social values have a great deal of ideological power in South Korea. Confucius regarded the role of the person with authority (e.g., ruler of a nation, teacher of a class, and parents in a family) as "an attractive model of what a person should be, like the polestar" (Kupperman, 2004, p. 107). The Confucian paradigm of "five relationships" refers to ruler/subject, father/son, older/younger, husband/wife, and friend/friend.

> In all of them but one—the relationship between friend and friend, assuming the friends are of exactly the same age, gender, and social rank—the relationships are unequal and require that the weaker party voluntarily submit to the stronger while the stronger exercises nurture and protection over the weaker. (Clark, 2000, p. 31)

Some anti-Confucianism movements have challenged the discourse on women in Confucian writing and thinking (Duncan, 2002); however, ageism, traditionally a significant aspect of Confucianism, has remained a strict custom for social conduct. As Hur and Hur (1993) wrote:

> One of the main doctrines of Confucianism is that respect is due to the elderly. In Korea it is clear, just by seeing them walk around, that the elderly feel pride and a kind of self-respect that is not seen in older people in Western societies. Many Koreans are shocked at how the elderly are treated in other societies. (P. 88)

Korean social relationships today are still vertical by age in Korean culture. Each person knows his/her position in relation to other people, individual preferences are not encouraged, and in fact, each person's emotions are to be suppressed because Confucianism stresses "the harmony of social relationships." For Korean teachers, the

Table 4.1 Faculty of Sarang School

Name (pseudonym)	Birth Year	Degree	The first year of teaching	Years of teaching experience	Class they teach
Jisu	1979	Two bachelor's degrees (engineering, early childhood education)	2007	13 months	Four-year-olds' classroom
Sumin	1986	Bachelor's degree (early childhood education)	2008	1 month	Three-year-olds' classroom
Hanna	1985	Bachelor's degree (early childhood education)	2008	1 month	Three-year-olds' classroom
Ayoung	1984	Bachelor's degree (early childhood education)	2007	13 months	Two-year-olds' classroom
Junglan	1982	Bachelor's degree (early childhood education)	2005	3 years	Five-year-olds' classroom
Heji (Ms. Song)	1977	Bachelor's degree (early childhood education)	1998	7 years	Four-year-olds' classroom
Yna (Ms. Lim)	1967	Bachelor's degree (social work)	1988	10 years	One-year-old's classroom
Sumi	1968	Bachelor's degree (early childhood education) Master's degree (early childhood education)	1989	9 years	Director

cultural belief that senior colleagues are wiser is powerful and shapes how beginning teachers feel they can act in their schools.

The school in which this study was conducted was an early childhood private school, located in the city of Mirae[1] in southwest Korea. The school served one hundred and twenty children, ranging in age from infants to five-year-olds. The staff included seven teachers and a director. Of these eight staff members, three, including the director, had more than seven years of experience; one had three years of experience; and four had fewer than two years of teaching experience. The data for this chapter, which was collected for a larger project, focuses on some of the teaching decisions made by three of the beginning teachers, particularly looking at the ways these decisions were impacted by their novice status in the school. Table 4.1 lists the faculty of Sarang School.

METHODS

This chapter was based on a larger study conducted by the first author, Su Kyoung, which examined how Korean early childhood education induction teachers perceived themselves as newcomers in the teaching world. Because a characteristic of case study is providing "a rich and thick description" (Merriam, 1998), this research typology was the best vehicle for providing "intensive descriptions and analyses of a single unit or bounded system such as an individual, program, or group" (p. 19). By using a case study approach, the researcher aimed to present an in-depth understanding of the situation, to describe the meaning for those involved, and to include descriptions of each participant's background and teaching contexts. The units of analysis for this case study were individual induction teachers who worked in a particular Korean early childhood education setting. Additionally, Su Kyoung engaged in cross-analysis of the individual cases (Stake, 2006). The cross-case analysis provided good opportunities to learn about complexity and contexts through the mediational experience of induction teachers.

The larger study focused on understanding induction teachers' lives at the intersection of social, economic, historical, and cultural contexts. Thus, the decision was made to select induction teachers working in the same school rather than selecting them from different schools. This allowed a deeper description of participants' lives in a natural setting. Using purposeful sampling (Merriam, 1998), four induction teachers working at "Sarang School" were invited to take part in the study, although only three of these teachers are discussed in this chapter. All the participants were women in their twenties, born and educated in southwest Korea. To understand the experiences of the individuals in this study, five primary data sources were used: (1) field observations, (2) focus group discussions, (3) individual interviews with participants, (4) field notes, and (5) photo essays created by the participants. The secondary data sources included the researcher's journal and artifacts such as e-mail messages, pictures, newspapers, school handouts, and lesson plans. The data were collected from April to August 2008. Each participant took part in three semi-structured individual interviews and two focus group discussions. The interviews and group discussions were conducted in Korean by Su Kyoung, a Korean. All observations were recorded with field logs and notes. All the raw data were transcribed first in Korean, and later translated into English. Each participant was provided with an initial draft of their case study written in Korean, so they could indicate any aspect of the narrative that made them uncomfortable in terms of inclusion in the study.

In this study, case analysis was integrated with narrative analysis (Manning & Cullum-Swan, 1994) and a grounded theory approach (Charmaz, 2006; Strauss & Corbin, 1998). In narrative analysis, the stories from the participants are both personal and social, reflecting a person's life history and the contexts where they live. In the present study, narrative analysis was based on the participants' photo essays, interviews, conversations, field notes, journals, and e-mail messages between teachers and Su Kyoung. In the integration of representational and presentational narratives, teachers' words or stories captured their thoughts, beliefs, and feelings. Additionally, a grounded theory approach, including constant comparison, open-coding techniques, and memo-writing, was used to guide case analysis. The findings were shared with the teachers for the purpose of member checking (Merriam, 1998).

In this chapter, we draw on the preceding data and analysis to explore the ways three beginning teachers negotiated particular challenges in their school contexts. The first case concerns two beginning teachers who found a certain schooling practice troubling, but were unable to make their concerns heard by their senior colleagues. The second case concerns a beginning teacher who succeeded in changing the school's practice.

HANNA AND AYOUNG: STRUGGLING TO CREATE OPENINGS FOR CHANGE

Each Monday morning all of the teachers and children at Sarang School would gather for an hour-long morning meeting,[2] which often included the singing of songs and the reciting of poetry. All the children are expected to sit and actively participate in the activities throughout this gathering time. However, the young children frequently found this to be a challenge, as demonstrated in the following anecdote taken from observation notes of an induction teacher, Hanna, and her three-year-old children.

> It is 10:30 a.m. on Monday. All Sarang's teachers and children gather in the school gym. Each classroom finds the place in which they need to sit under the direction of each classroom teacher. Hanna's classroom children sit in the first column from the right, and Hanna sits in the back of the last row. One of the experienced teachers, Ms. Lim, who has been teaching for about 10 years, jumps on the stage and starts the national anthem holding a microphone in her right hand to the melody of the piano, which one of the new teachers plays. Hanna and her children sing the song. And then Ms. Lim keeps teaching songs for all the children on the stage and the children sing along. One of Hanna's children comes to her and says, "I want to go to the potty." Hanna hurries to take her to the restroom in the hallway. With a smile, Ms. Lim asks the children, "Who wants to come here to sing this song in front of your friends and teacher?" Some children raise their hands and the teacher points out the children with her finger. Most of young children, including Hanna's children, are not paying attention to the activity. In particular, the infants and the toddlers start going around the gym, and their teachers prevent them from moving around. After the children come in the front, they sing a song and Ms. Lim encourages the other students to clap their hands. At that time, Hanna comes in the gym again with the girl she took to the restroom. After the singing time finishes, Ms. Lim gets down from the stage and goes to her children. And the director jumps on the stage and sits on the chair which is located in the center of the gym stage. The director introduces a poem to the all children. She says, "Repeat after me, please." When the director reads the poem, the children and the teachers repeat it. Hanna's children keep moving and twisting their bodies here and there. Some children are talking with the next child without focusing on what the director is doing. Hanna goes to the children and makes a quiet sign with her pointing finger. She looks nervously at the director and the other teachers continuously. (Researcher's field notes, April 12, 2008)

Hanna described the morning meetings as "meaningless" and said that her children would benefit more from engaging in free play during this hour. She wrote about this issue in her photo essay:

> My classroom children are three-year-olds. It is hard for them to stay for one hour in Monday meeting. As a member of Sarang School, I

understand that it can be a good time for all the children and the teachers [to meet as a group]. However, it's too much for my classroom children. They don't understand why they need to be there. And sometimes the songs to learn there seemed very difficult for this age. When my children were not involved in the activity, I felt nervous in that I was responsible for their behavior. I felt other teachers thought of my class as a terrible class. When my children asked me not to participate, I could not say anything. What should I do? Should I listen to my children's words? Or should I comply with the school policy without saying anything like the other teachers? (Hanna's photo essay in the group discussion, June 26, 2008)

In the photo essay, Hanna expresses a concern about what the other teachers think of her classroom when her children do not sit still in the morning meeting. Hanna expressed a desire to use the morning meeting time for free play, which she believed was more important for understanding children's perspectives and developing a sense of community than participation in the whole school gathering.

When Hanna had the opportunity to engage her children in play, she was able to focus fully on them, encouraging their imagination and verbal expression. One morning she pretended to cook dinner with some of her children in the kitchen. As one child pretended to wash dishes, she called out: "O.K. It's done. Mom, take the food out of the oven." Hanna reached for the pan and then stopped, shouting: "It's very hot. What can I use for it?" The child told her to use an oven mitt. Hanna put the food on the table and said, "Smells good. I'm ready. Let's have dinner." Then the girl whispered that they were not ready to eat because they were waiting for a guest.

In this episode, Hanna interacts confidently with her children, scaffolding the little girl's thinking and encouraging her to talk. Hanna demonstrated through her actions, both during indoor and outdoor free play, that her desire to give the children more of this unstructured time was not based on a wish to take it easy in the classroom or to be freed of the responsibilities of teaching. However, her vision of what was most productive for her children was very different from that presented by the director and all the experienced teachers in the school. As a result, although Hanna believed that her classroom children needed to interact with her through free play rather than to passively sit through the morning meeting, she as a (relatively) new teacher felt compelled to comply with the school ritual and enforce her children's participation in the event.

Ayoung, another first-year teacher, expressed similar concerns about the morning meeting. Without an assistant to help with the six children, Ayoung's two-year-olds were often restless during this meeting as shown in the following excerpt from observation notes:

> When the entire group is singing in the Morning Meeting, suddenly, one of Ayoung's students stands up and runs toward the stage. After following and catching him, she takes the boy to the mat and sits him down. Most of Ayoung's students don't pay attention to the meeting and struggle to escape from the mat. Ayoung just keeps watching her students so that they will not run away. And she says to her classroom children, "(Whispering) it's almost finished. Please, sit down. Don't move." Nevertheless, the students try to move repeatedly while Ayoung prevents them from moving. (Researcher's field notes, April 12, 2008)

Like Hanna, Ayoung believed that her children would benefit from engaging in free and pretend play in the classroom, rather than sitting through the whole school meeting. She believed that some of the other teachers including the experienced teachers shared her opinion, and that open discussion among the staff could create changes that would benefit the children. However, Hanna found that her colleagues, who shared her concern about the morning meeting, were not interested in taking action for change.

> If everybody is silent, we can not expect improvements for this school. When I said to the director that I don't want to take part in the Monday meeting with my students, she just answered that the meeting is to create unity for every Sarang student and objected to my opinion. Is it possible for one and two year old children to have a sense of unity? Anyway…after that, I have not mentioned it any more. Everything is the same. No discussion here. No opinion considered here. (Interview, May 21, 2008)

Here, Ayoung reported an unsuccessful effort to introduce a change in the school that she believed would be helpful for her classroom children. However, she was not able to muster support or interest among her colleagues. The faculty meeting that Ayoung envisioned was one where all teachers would have a chance to talk, regardless of their status. However, she felt that she had primarily experienced only one-sided communication. It seemed to her that only the experienced teachers spoke, and the new ones just listened. When novices spoke, the more senior teachers often looked away or ignored what was said. Comments by novices were much less likely to be taken up by the group than those by more experienced teachers. Most teacher meetings often looked like the one described in the following extract:

> Today, after cleaning up their classrooms, the three, four and five year old classroom teachers have a faculty meeting in the office. Ms. Song, who is the most senior teacher in this meeting with seven years of experience, starts the meeting and says, "We need to fill out the student record forms by this Friday at 6:00 p.m. You need to include the date, health information...You need to write down what you know. From the next year on, we will report more specifically." Other teachers are listening quietly. Ms. Song keeps talking, "Let's talk about the preparation for the summer camp." Other teachers are still listening and taking notes. Ms. Song asks (looking at the clock), "Do you have any special issues today?" Nobody answers. Ms. Song keeps speaking, "Oh, I forgot about writing something for the parents. You need to write something about the child for the parents everyday. The director says we need to write as much as we can." During the meeting, Ms.Song keeps giving information to the other teachers. With their heads down, the other teachers are taking notes. All the teachers including the experienced teachers who are participating in the meeting look very serious and unhappy. (Researcher's field notes, June 24, 2008)

Despite their desire for change, neither Hanna nor Ayoung ever raised the issue of young children's participation in the morning meeting at a teacher meeting. Both said that the pressure to remain silent and to listen to the experienced teachers was too much for them to overcome because these morning meetings were created by the experienced teachers and the director without any input from the novice teachers. The experienced teachers unanimously supported this decision.

SUMIN: A NOVICE PUSHING FOR A CHANGE

Although Hanna and Ayoung were not able to remove their children from the morning meeting, we do not wish to argue that novices, even in expert-dominated contexts, can never make changes in the larger organization. This section examines Sumin's efforts to change the way Sarang School handled certain parental involvement practices. Like Ayoung and Hanna, Sumin was also a first-year teacher. She was responsible for four-year-olds. In interviews, Sumin said that she believed parental involvement could be a good channel for building positive relationships and energetic communications with families. However, she sometimes felt at odds with other Sarang teachers over this issue. In an interview, she explained:

> The director and all the experienced teachers did not agree with the parents' participation originally. For example, some of parents in my classroom really wanted to take part in the school's program as volunteers. In the case of the "Sarang Flea Market" play, they learned how to interact with the children and how to prepare the activity because they had opportunities for observing the children and teachers. They were also able to understand the educational philosophy of this school. Actually, in the school's online homepage, parents often wrote, "This school needs some activities for the parents." They wanted to be helpers for the teachers as well as the children. They had good intentions as well. You know, we are the members of the same community. But some teachers didn't like the participation of the parents. I did not understand them. (Interview, May 28, 2008)

Unlike Ayoung, who raised the issue of removing her children from the morning meeting only once, Sumin repeatedly raised questions about the parental involvement policy. When she first shared her thoughts with the experienced teachers, they responded negatively. And next, when she mentioned her ideas to the director, she got the same response. However, she constantly maintained her opinion in faculty meetings. She commented:

> This school needed to change its views about parent involvement. The process was...First, when I proposed the idea about parent involvement, they (the experienced teachers and the director) ignored me. And when I put it on the agenda in the faculty meeting, of course, I was afraid of their criticism, but I was brave. It seemed they worried that parents would criticize the teachers after participating in the program and observing the school. I felt that they were cowards at that time. You know, it is important to show how hard we work here to the parents and their help can be very educational for the children. If they criticize us in a negative way, we need to listen to them for the school improvement. However, we don't need to think that they are just evaluators. Furthermore, they would not just complain because they think that they are also members of this school. (Interview, May 29, 2008)

When Sumin proposed the participation of the parents for a school event, "Sarang's Flea Market," in the faculty meeting, she felt that the atmosphere of the meeting became chilly. However, she became an advocate for it and finally the other new teachers agreed with her in the middle of the long meeting. Sumin stated that many parents were interested in "Sarang's Flea Market," and that some of them really wanted to participate. Also, she explained the value of this kind of parental involvement in the school. Yet, some teachers and the director thought parents could criticize the school management, curriculum, or instruction if the parents were given this opportunity to observe the school. In the end, the experienced teachers and the director grudgingly accepted Sumin's idea, adding, however, that they probably would not repeat this experience because parental participants could be challenging for the teachers. Nevertheless, all the teachers as well as the parents were satisfied with the results of the parental involvement. Sumin felt great confidence at that moment as a teacher, noting:

> When I proposed the parent involvement, you guys know, my face turned red because it was the first time I gave my opinion as a new teacher in a formal meeting. Anyway, I survived the moment. The result made me feel confident as a teacher. Also, I proposed one more idea at that time. I said we needed to give an evaluation form to the parents after participating in the activity. I did not see the evaluation results yet because only the director and the more experienced teachers shared them. (Group discussion, June 26, 2008)

Sumin's anecdote demonstrates that with determination and bravery, a novice in this environment could create some change. However, it took her a great deal of effort to make the change, and even after her hard work, she still was not included in the initial meetings about parental feedback from the event nor was her idea incorporated into the standard practice of Sarang School. As a novice she was able to create change in a particular practice, but unable to unsettle more deeply held beliefs about whom had a right to be at the table and to have access to information when decisions were being made.

DISCUSSION

In South Korea, some educators use the word "cho'zza[3] teachers" or "baby teachers" to describe novice teachers. Sumin, Ayoung, and Hanna each had to negotiate their teaching decisions in response to this label. In a discussion between the participants, Hanna defined a "cho'zza" teacher as "a lonely teacher who had to survive without anybody's help and as a silent teacher who could not speak out without restraint" (Group discussion, May 13, 2008). Whenever these teachers participated in the dialogues within the school context such as conversations with experienced teachers or parents, they considered themselves early childhood induction teachers and, as a result, faced challenges in terms of their teacher identity and their ability to act in ways they thought were appropriate. In particular, Ayoung and Hanna accepted that they would need to attend morning meetings with their children when the director articulated that this was important. They did not feel empowered to challenge this idea further despite their belief in the importance of free and pretend play for young children. While Sumin did continually lobby for greater parent involvement, because of her "cho´zza" status, she too accepted that others would first see the parent evaluations of the activity she had been so central in organizing.

The challenges to the beginning teachers stemmed from conflicts between their own educational belief systems and the actual authoritative discourse they had to negotiate as novice teachers. Their positioning as novices not only gave Sumin and Ayoung less credibility when speaking in public, but it also diminished their sense of their own right to call practices into question. Whenever they were recognized as inexperienced teachers, they had to rethink their identity as teachers and their options for articulating their beliefs.

In their role as new teachers, all the participants in this study experienced certain traditions and habits that were imbued with subtle power relations between faculty members. Sumin, Hanna, and Ayoung were autonomous teachers who were capable of carrying out many things by themselves in the classroom, but they also had to play a passive role in the hierarchical teacher community. Their status as newcomers influenced their roles in the teacher community: doing chores, respecting and listening to experienced teachers, and, oftentimes, being silent. In an interview Ayoung mentioned that deferring to more experienced teachers was considered the "natural" way for new teachers to act in her school. The experiences of these three teachers suggest that induction teachers might want to find their own way by themselves rather than rely on the help of another teacher, which might cause them to feel coerced rather than empowered. They do not want to get instant advice from others but to discuss their difficulties with others. Hanna explained that every classroom had a unique context that only the classroom teacher could understand and manage. She felt the experienced teachers tried to give her advice because they thought she was a baby teacher. She stressed, "The final decision was made by me, not by them." These examples show that more experienced teachers, who might be expected to act as mentors, even with their good intention do not always act in ways that support induction teachers.

In Lave and Wenger's terms (1991), the beginning teachers experienced themselves as legitimate peripheral participants. However, rather than accepting the peripheral participant role as legitimate, creating change in the school required that the beginning teachers act in ways that might not have been considered developmentally appropriate within the context of the community—such as through raising concerns to the administrator or speaking up repeatedly in staff meetings. When Sumin refused to let the issue of parent involvement in the play drop, she resisted the imposed criteria and status of "novice" in order to represent herself and the children and their families. In this case, a positive outcome—that of greater parental involvement in the school—resulted from Sumin's act. This is an example of an induction teacher introducing a new pedagogy and resisting the dominant thinking about schooling offered by experts. In this case, Sumin's learning could be described not as moving her closer to the center of the existing community, but as learning that she had the power to resist community norms and traditions and push toward a new center.

As Sumin's case illustrates, in terms of learning and teaching, the induction teachers could sometimes position themselves as resources, rather than as apprentices or novices in the schools. They wanted to collaborate with experienced teachers and to have their new ideas accepted in the school community. Sumin's story challenges the idea that novice teachers move in *a linear way* toward becoming expert teachers and raises questions about theories that position induction teachers as "peripheral" participants in school communities. These teachers can be seen as different from experienced teachers with particular perspectives and experiences, but not as necessarily moving toward a "better" stage in teaching. In this study, Sumin believed that all the teachers needed to have open minds about parental involvement for the sake of the students. Hanna and Ayoung believed that children needed to be actively engaged in play, rather than passively watching the morning meeting. These goals, which many university educators would praise, might have been easier for the induction teachers to achieve if they had not been so consistently labeled as novices who had much to learn from, and nothing to teach to, the experts in the building. This positioning occurs not only in schools, but also in research literature that focuses on the struggles of induction teachers to the exclusion of the strengths and ideas they bring to their new schools. Future research needs to consider that the struggle faced by induction teachers is sometimes not the result of their own difficulties in adapting to a new environment, but an intelligent, principled response to an environment that needs to change.

From a Bakhtinian perspective, Sumin, Hanna, and Ayoung can be seen as acting from their unique places in Being and learning through their own interpretation of others' actions and ideologies. For example, as Sumin worked toward including parents in the play, she gained a sense of self through her engagement with those around her. She confirmed her beliefs about the importance of including parents, but also learned that she could be brave when necessary. Thus, she does not need to be seen as moving closer to the center of the existing community in order to learn, but can be seen as learning through her active and unique resistance to ideas and discourses with which she disagreed.

Similarly, although Hanna and Ayoung were less effective at creating change in the school, they could be seen learning as they positioned their own unique Beings in relation to those around them. Although the practice of morning meetings did not change, the opportunity to disagree with the practice, both privately and publically, provided a chance for Hanna and Ayoung to learn about the kind of teachers they wanted to be. The confrontation with morning meetings caused them to articulate a belief in the importance of play in young children's lives, to find time for that play when they could in the classroom, and to learn to make an argument in favor of it to the researcher. At the end of the year, Ayoung considered leaving Sarang School, in part because she could not make her own ideology about teaching fit easily with the ideology of the school. This too can be seen as an act of learning, as Ayoung begins to articulate and consciously choose what kind of teacher she will be.

In addition to devaluing the strengths newcomers bring to the profession, we want to argue that conceptions of teacher learning and growth that rely on developmental views that see new teachers moving from outsiders to central participants in a community provide too little space for conceptualizing educational change. If induction is seen as a stage that teachers must move through, where they leave behind the immature and impractical conceptions they developed in their preparation programs and toward the conceptions and practices of the experienced teachers in the existing school, then there is little hope that schools will do anything other than sustain current ways of thinking and teaching. However, if we could begin to conceptualize induction teachers as community members who equally contribute to and benefit from the community, then new possibilities for changing school culture could be mapped out. Further work on induction teachers needs to examine the power relations between newcomers and oldtimers, the possibilities for learning for and from all members of the community, and the strengths and weaknesses that all teachers bring to the table. By conceptualizing induction less as a unique stage that is plagued by difficulty and "growing up" and more as an opportunity for changing a community, researchers could begin to open possibilities for both beginning teachers and schools.

NOTES

1. Names of places and people in this study are pseudonyms.
2. Sarang School has a morning meeting for moral instruction and musical practice for one hour every Monday morning. All children from six-month-olds in the infant classroom to six-year-olds in the oldest children's classroom gather at the school gym. The children sing the Korean national anthem and a new song of the week, and learn children's rhymes and finger plays together. The director and experienced teachers who take turns lead this morning meeting in front of the children and the other teachers.
3. Originally, this word is from the Chinese characters, 初 (beginning) 者 (person), with the pronunciation, "choza." Korean people usually say this word with a strong accent, "cho′zza," signifying a degrading meaning. This term means "a person in the first [beginning] step" or "a newcomer." Usually, when Korean people use this term, the context is when a person is looked down upon as a beginner with a lack of ability or skill. This expression is similar to "new kid on the block" in English. It carries negative connotations and would not usually be used in front of novice teachers except for situations of telling jokes or describing a beginner to others. At times, the novice refers herself as "cho′zza" in a self-deprecating manner for an excuse or humor.

REFERENCES

Bakhtin, M. M. (1993). *Toward a philosophy of the act* (V. Liapunov, trans.). Austin: University of Texas Press.

Berson, M. J., & Breault, R. A. (2000). The novice teacher. In B. E. Steffy, M. P. Wolfe, S. H. Pasch, & B. J. Enz (eds), *Life cycle of the career teacher* (pp. 26–43). Thousand Oaks, CA: Corwin Press.

Bloom, R., Sheerer, M., & Britz, J. (1991). *Blueprint for action. Achieving center-based change through staff development.* LakeForest, IL: New Horizons.

Brock, B. L., & Grady, M. L. (1998). Beginning teacher induction programs: The role of the principal. *Clearing House, 71*(3), 179–183.

Charmaz, K. (2006). *Constructing grounded theory: A practical guide through qualitative analysis.* London: Sage.

Clark, D. N. (2000). *Culture and customs of Korea.* Westport: Greenwood Press.

Costigan, A. (2004). Finding a name for what they want: A study of New York city's teaching fellows. *Teaching and Teacher Education, 20*(2), 129–143.

Duncan, J. B. (2002). Uses of Confucianism in modern Korea. In B. A. Elman, J. B. Duncan, & H. Ooms (eds), *Rethinking Confucianism: Past and present in China, Japan, Korea, and Vietnam* (pp. 431–462). Los Angeles: University of California.

Feiman-Nemser, S. (2001). From preparation to practice: Designing a continuum to strengthen and sustain teaching. *Teachers College Record, 103*(6), 1013–1055.

Flores, M. (2007). Navigating contradictory communities of practice in learning to teach for social justice. *Anthropology & Education Quarterly, 38*(4), 380–404.

Hebert, E., & Worthy, T. (2001). Does the first year of teaching have to be a bad one?: A case study of success. *Teaching and Teacher Education, 17*(8), 897–911.

Horn, P. J., Sterling, H. A., & Subhan, S. (2002). Accountability through "best practice" induction models. Paper presented at the annual American Association of Colleges for Teacher Education, New York, NY. (ERIC Document Reproduction Service No. ED 464 039).

Howe, E. R. (2006). Exemplary teacher induction: An international review. *Educational Philosophy and Theory, 38*(3), 287–297.

Hur, S. V., & Hur, B. S. (1993). *Culture shock! Korea.* Portland: Graphic Arts Center Publishing Company.

Katz, L. (1972). Developmental stages of preschool teachers. *Elementary School Journal, 73*(1), 50–54.

Kupperman, J. J. (2004). Tradition and community in the formation of character and self. In K. Shun & D. B. Wong (eds), *Confucian ethics: A comparative study of self, autonomy, and community* (pp. 103–123). Cambridge: Cambridge University Press.

Lave, J., & Wenger, E. (1991). *Situated learning: Legitimate peripheral participation.* New York: Cambridge University Press.

Lortie, D. (1975). *Schoolteacher: A sociological study.* Chicago: University of Chicago Press.

Manning, P. K., & Cullum-Swan, B. (1994). Narrative, content, and semiotic analysis. In N. K. Denzin & Y. S. Lincoln (eds), *Handbook of qualitative research* (pp. 463–477). Thousand Oaks, CA: Sage.

Merriam, S. (1998). *Qualitative research and case study applications in education: Revised and expanded from case study research in education.* San Francisco, CA: Jossey-Bass.

Metzler, M., Lund, J., & Gurvitch, R. (2008). Chapter 2: Adoption of instructional innovation across teachers' career *stages. Journal of Teaching in Physical Education, 27*(4), 457–465.

Morrell, E. (2003). Legitimate peripheral participation as professional development: Lessons from a summer research seminar. *Teacher Education Quarterly, 30*(2), 89–99.

Sabar, N. (2004). From heaven to reality through crisis: Novice teachers as migrants. *Teaching and Teacher Education, 20*(2), 145–161.

Stake, R. E. (2006). *Multiple case study analysis.* New York, NY: The Guilford Press.

Stansbury, K., & Zimmerman, J. (2000). *Lifelines to the classroom: Designing support for beginning teachers.* San Francisco, CA: WestEd.

Strauss, A. L., & Corbin, J. (1998). *Basics of qualitative research: Techniques and procedures for developing grounded theory* (2nd ed.). Thousand Oaks, CA: Sage.

Veenman, S. (1984). Perceived problems of beginning teachers. *Review of Educational Research, 54*(2), 143–178.

Wayne, A. J., Youngs, P., & Fleischman, S. (May 2005). Improving teacher induction. *Educational Leadership,* 76–78.

Wenger, E. (1998). *Communities of practice.* New York: Cambridge University Press.

CHAPTER 5

Responsivity rather than Readiness

M. Elizabeth Graue

I believe that education, therefore, is a process of living and not a preparation for future living.

—*Dewey, 1897*

My family is in the process of planning a trip, complete with airline schedules, hotel reservations, car rental, and maps. This is a new thing for us—in past years we have rented a lake house, planted ourselves for two weeks, and refused to schedule anything. The timeless quality of these lake vacations was a remedy to over-scheduled lives with too many deadlines. Our days were filled with books, cherry pie, sand, and naps. But this year will be different. We have a plan. It came about in part because my children had started to ask why we never went anywhere (which makes me wonder about their sense of place) and in part because of my own wanderlust. I've engaged a travel agent, trolled websites, and purchased guidebooks. We will be busy, on the move, balancing experience moment-to-moment with thinking ahead.

While the process of this vacation is different, I've come to realize that the intention is in many ways the same—to change our context and rhythms so that we change our relationship to the world. Will one approach be inherently better than the other? I hope not as I recognize that the value of an experience is in relation to its goals. While my traditional vacation approach might seem to lack goals, it is in fact designed mindfully with the goal of changing my relation to the clock, to relearn the value of just being. My upcoming vacation is also designed to challenge that relation by plopping me down on another continent and making me live on a different timetable. In removing me from my everyday context, I will again get oriented to the here and now.

By now you might be wondering what my vacations have to do with this book—it's not a travel guide or memoir. It is focused on education rather than on recreation. But I hope to draw a connection through an exploration of the ways that we orient our action. Should we be motivated by the here and now, the past, or some future goal? In both the case of my vacation and in the discussions presented in this section, this question is foundational. I will use the opportunity of this commentary to converse with the authors of this section and to add my own perspective on a variety of themes.

> White Rabbit: [*singing*] I'm late / I'm late / For a very important date. / No time to say "Hello." / Goodbye. / I'm late, I'm late, I'm late.

One of the most powerful themes in *Alice & Wonderland* is the notion of time. The upside down world of Wonderland forces a rei-magining of time and helps illustrate how arbitrary our use of time is. In a curious way, this collection of papers serves the same functions, helping us to see the unexpected assumptions, practices, and implications of commonly held educational ideas. Each of the authors framed a critique about how education engaged visions of the past, present, and future—in relation to readiness, teacher induction, conceptual-izations of ADHD, and construction of the middle school curriculum. They illustrate how particular notions of temporality are woven into our conceptions of childhood and our warrants for teaching.

Beginning in utero, children are marked by time, understood in relation to a nine-month gestation, number of months since birth, being part of an age cohort at school. They might be framed by the calendar and called a "summer birthday" or described as an "early reader." The clock always seems to be ticking, locating children within a set of temporal norms. While anthropologists have helped us see that many of these norms are *cultural* rather than physical (Rogoff, 2003), they press upon our understandings of children to such a degree that they create a child, and through that creation, a set of rules and roles for the adults around them.

Some examples of how time shapes our practice—we give the gift of time (delaying school entry, in a less than neutral metaphor) to children who are seen as immature while early intervention is sug-gested for children whose development is delayed or who are from families living in poverty. Both immaturity and delayed development are characterized by having progressed less than expected. So are these opposite treatments for similar symptoms? What typically distinguishes the candidates is socioeconomic status, with the former—delaying school entry—chosen for middle and upper class children, typically boys, and the latter—early intervention—chosen for working and lower class children. For the affluent, protection from the demands of schooling is seen as appropriate because their homes are assumed to be rich sources for learning and for those in poverty, a head start is deemed necessary because their familial contexts must lack the resources to produce well-prepared children. The view of time and the resources available to leverage good outcomes come together to construct particular visions of children and responsibilities for adults.

THE BOGEYMAN NEXT YEAR

A very specific kind of manifestation of time permeated these chapters and much of educational practice today. It might be viewed as teaching to build a foundation for future learning, an irrational fear of something in the future, the imprint of backmapping from standards. Lilian Katz and Sylvia Chard (2000) call it favoring the vertical relevance of education rather than the horizontal. When we favor vertical relevance, we justify what we do in terms of future payoff or perceived requirements rather than rationalize it in terms of current needs. This kind of logic plays out when we shorten free play time in preK/K because we think they won't have it in the K/first grade. Or we require students to use assignment notebooks in fourth grade because we think they will need to use them in middle school. Or we expect eighth graders to keep every paper generated in a history class and place them chronologically in a binder for the semester because we think it helps teach organization for high school.[1]

It is well illustrated in a couple of the chapters in this section, at different levels of schooling. That future bogeyman was well captured by Hilary Conklin's call for mindfulness in middle school teaching and learning. Her chapter reinforced for me the parallels between early childhood education and middle school education, something that came alive to me as the parent of a middle schooler. In my older son's middle school experience, I was struck by the construction of middle school as a special time, designed to address the particular needs of pre- and early adolescents. The logic was quite similar to the traditional kindergarten that was seen as a transition between home and school focused on the physical, social, emotional, and cognitive

needs of five-year-olds. In much the same way that this "special" status has waned as kindergarten has melded into the elementary program, middle school (with the exception of sixth grade) seems
frighteningly similar to high school.

Conklin's critique of the level of intellectual engagement in middle
school points to the infantilization of middle schoolers who are hit
with a double whammy—they are seen as intellectually unable to
engage in "real" thinking yet they are forced to endure many types of
preparative activities in the name of future educational experiences.
As a result of this diminished view of their students, the teachers are
limited in the types of curriculum they might engage in, stuck
between a view of addled preadolescents and fear of the demands of
high school. Relying on notions of mindfulness, Conklin reorients
the goals and intentions of the middle school curriculum in apprecia-
tion of students in the here and now. If middle school went on vacation, would it go to the beach or a scheduled road trip? Conklin
wants us to work on being in the moment whatever trip we take.

The bogeyman played out in the Parks and Bridge-Rhoads description of the implementation of the *SRA Language for Learning* curriculum in a preschool, through what they call a forward-looking
curriculum. According to the program's authors, some students must
learn the language for learning in school rather than at home. The
curriculum explicitly teaches the practices deemed necessary for success. This includes labeling and speaking in full sentences, two practices that were eerily familiar from days in catechism class. Interestingly,
the curriculum's orientation to time is as backward leaning as it is
forward looking in its assumptions about the language experiences of
the students in life prior to preschool. The students were sandwiched
between a history that was deemed inadequate and a future full of
academic bogeymen.

In an interesting turn the authors argue that using the notion of
developmentally appropriate practice is inadequate to interrupt the
logic of the L for L curriculum. Their line of reasoning is that if you
suggest L for L is not appropriate for four-year-olds, then it must be
good at some later period of a child's life. That struck me as being
similar to saying that if you suggest marshmallows are not good for
breakfast, then you are implying that they are good for dinner. As
someone who never wants to eat marshmallows, that logic worries
me. But more importantly, I think it overstates the implications of
thinking about developmental appropriateness. Let me give a counterexample. Most people would say that teaching calculus[2] in kindergarten is not developmentally appropriate but that doesn't mean that

all children should learn it at a later date. It just means that given the individual development of most five-year-olds, the typical expectations for development of children that age, and the cultural relevance of the content, it's not a good bet.

WHO'S AFRAID OF THE BIG BAD DEVELOPMENTALISM?

Several of the chapters take direct aim at the notion of developmentalism, framed as a way of conceptualizing children and their education. In some cases, developmentalism is a bad guy, a concept that causes us to midjudge children in the here and now and that moves caregivers, teachers, and parents to view children in terms of their future stages. I must confess that this is a view that I have found compelling in the past when I described the readiness related actions of preschool and kindergarten teachers and parents. Much of the work was oriented toward what I called the *kindergarten prototype* (Graue et al., 2003), a conceptualization of appropriate development for successful kindergartners that was shaped by the normative practices of home and school. I recognize that it has been argued that notions of progress made developmentalism possible but wonder if it is an overstatement to think of developmentalism as narrowing our views of children solely to the future.

A compelling counterexample is given in the Park and Parks chapter on novice teachers in Korea. In their retelling of Hanna and Ayoung's frustrations with an extended morning meeting for preschoolers, the young teachers wished that their three-year-olds could spend their time in free play instead of the large passive group meeting. Although not stated as such, their logic for not including the younger children in the morning meeting read as if it came out of an argument for developmentally appropriate practice, focused on the developmental needs of their students. In this case, developmentalism was a progressive logic, something that grounded action responsively in the here and now. The authors suggest that the teachers are acting out of an internally persuasive discourse, and resisting the authoritative discourse of the school. However, in Bakhtinian terms (1982), these teachers were resorting to a *different* authoritative discourse than their more senior colleagues, that of developmental appropriateness. This reminds us that the dominant or authoritative discourse is always relative—in one case it is a hegemonic force, in another it is seen as a more equitable conceptualization of children's needs.

It may be that we would be better off resorting to critiques of the *uses* of developmentalism rather than of the construct itself. I know that this is a bit like the old argument that people not guns kill people, but precision in this kind of critique is important. This is the tack taken by Lee in her chapter that reframes child development from a cultural psychological perspective. Through a description of the development of developmental psychology, Lee shows that while the field has grown into a framework that is systems oriented, its use in early childhood education remains stunted in a linear conceptualization. This impoverished notion requires development to come before learning, cementing a readiness perspective into the actions of educators. In her analysis of understandings of ADHD, Lee deftly illustrates the problems with this approach, which conflates child characteristics, school norms, and the pathology that is represented by being labeled as ADHD. In a number of the cases shared, teachers spoke about behavior in relation to norms that they suggested were developmental when, in fact, they were describing norms that were social or cultural. The utility of cultural psychology, which conceptualizes culture as a key developmental context, is shown clearly in Lee's illustration that ADHD comes to be in cultural contexts such as schools.

> As you know, you go to war with the Army you have. They're not the Army you might want or wish to have at a later time. (Donald Rumsfeld as cited in Kristol, 2004, P. A33)

One pervasive implication of forward-looking practices that ignore the present is that educators come to *prefer* an iconic child with particular skills, with suitable maturity, often justifying it by stating perceived demands of the future. More than escalating curriculum or favoring vertical relevance over horizontal relevance, these practices erase the child the curriculum was originally designed to address and replace it with an easy-to-teach student. This child can sit on the rug during morning meeting, doesn't work on tying her shoes during the reading of a book, knows exactly how far to take silliness, and can engage in intellectually challenging academic content. The problem with this is that it crowds out not only the "typical" child, but it also eclipses the actual children in the room. It enacts the problem of not being present in the moment with children. It also implies that there are children that cannot be taught, children who are not ready for your pedagogy. In my one moment of agreement with Mr. Rumsfeld, I would make a parallel argument—you teach the children you have, not the children you want or hope for at a later time.

Some might blame accountability and the standards movement for this forward motion in the life world of the classroom, but this displacement of the here and now from education has been in place for as long as I can remember. It is a key dimension of exclusionary readiness practices, where children are not allowed into early elementary grades because they are [fill in the blank—too small, too male, too close to the school entrance cutoff, too busy, too immature].

AGENCY

What these four chapters remind me is that teaching is a practice that has incredible potential for agency. Educators are presented with a variety of rationales for action and goals toward which they work. Each of these chapters highlight problems of disconscious practice; acting without an awareness of the tools used for action. Each analysis highlights the potential of other ways of thinking and acting, typically through alternative theorizations. But the very act of suggesting alternatives heightens attention to possibility for agency and a repositioning of the individual and the community.

Two Bakhtinian constructs are helpful in this repositioning. The first, *answerability* (Bakhtin, 1993), is a kind of ethical responsibility. Answerability poses lives as an unending series of consequential choices—as a result, it is a response in life's dialogue. When life is conceptualized as a response, but more importantly, as an ethical response, the interactions between teachers and students take on new meaning. The ethical possibilities and consequences of instructional practices become the core of education decision-making. In considering the chapters in this section, answerability might be used to think about *how* participants make decisions for constructing curriculum or the role of teacher or child. At a different level, answerability could be used to examine how researchers conceptualize the actions and motives of their research participants and how they depict a context to further their goals.

The second, *addressivity* (Bakhtin, 1986), is based on the idea of dialogism and conceptualizes life relationally. When life is viewed dialogically, it is seen as a response, a link in a chain of responses. From this perspective, all action emanates from a specific time and place. But equally important, action is posed to and for equally specific locations and purposes. Our lives are not random, nor are they totally determined. They are constituted by a history that shapes what has been and what is. Further, these histories focus on how we interpret the world and the future that we create. Our actions have trajectories, set within specific contexts and oriented to particular ends. Addressivity depicts a sense of culture and agency, a sense of life as both determined and open.

> Nothing means anything until it achieves a response. (Holquist, 2002, P. 48)

Bakhtin's dialogic perspectives on lived experience locate the individual within a social and historical world filled with interactions and responses. Answerability and addressivity constitute each other—the social is individual and the individual is social (Graue et al., 2001). Locating these constructs in the genre of education asks us to consider the historical meanings attached to schooling and to anticipate how our actions respond to particular individuals, groups, and goals. If we consider teaching as an ethical act, our actions are not just pedagogical, they are moral. As such they should recognize our historical positioning and address responses in a particular trajectory for specific ends.

DEVELOPMENTALLY RESPONSIVE PRACTICE

What kind of practice will be most likely to support both teachers and students in today's schools? What tools do we need to be answerable in our actions in ways that recognize our particular addressivity? I suggest that we might actually use a construct that has been critiqued in this book but that can serve to provide some traction if reconceptualized. The notion of development is a key thread in educational practice (particularly in early childhood) and something that is understood by diverse constituencies. Repurposing the idea of development, capitalizing on its historical importance but working to negotiate its application in a new context could provide good leverage for rethinking teaching and learning.

The approach that I suggest uses the notion of development to signal attention to past experience, present needs, and future learning. This conceptualization is relevant if we work from the assumption that teaching should recognize what children bring to school, their specific needs on a given day in a given place, and the goals we have for their experience. While some might argue that this signals a yearning for later learning, repurposing development with attention to the here and now reorients our sightlines from some later stage to enriching life right today, in this classroom space, while being mindful of the future.

I'd also like to highlight the fact that development, rather than being a process of the individual, is an accomplishment through the cultural context. From this perspective, it is very difficult to ascribe "problems" in classrooms solely to the student. Instead, problems are co-constructed; they help us understand why things might deviate from the patterns we expect. But problems are generated when the patterns become more visible than the individuals, when our expectations take on more salience than our interactions. What makes a new way of thinking about development helpful is a rebalancing of how we use our expectation. If a prototypical student is what guides all teaching, then we're missing today's student in all his/her glory. If we do not expect that middle schoolers can think and learn and interact in ways that are complex and important, it is not likely that we will develop curriculum that promotes that kind of interaction. If we conceptualize children living in poverty as less than more affluent peers, then that is what we will see. Yes, there are problems in education, problems that individual teachers have to deal with in individual classrooms. But if we focus on what is not there rather than what is, if we orient toward some generic norm rather than our own students, we are missing the whole point of teaching.

I suggest developmentally responsive (Graue et al., 2003) rather than developmentally appropriate practice to highlight the active and answerable nature of teaching. The notion of responsivity is key to constructing education practice that locates instruction within moment-to-moment interactions between teachers and students. It focuses on the contingent nature of instruction—that practice needs to be oriented between a student and a pedagogical goal.

It might seem that the forward momentum of much education practice is a recent phenomenon, a product of accountability and competitive forces. But the question of what, when, and why we teach is classic. It has been an element of discussions in education for as long as we have designed formal activities for children. It was described quite nicely by Lilian Katz (1973) over thirty-five years ago in her distinction between teaching and performance. She was concerned about:

> Teachers who do what they do in order to please a third person who is not even there. Often they say that what they're doing, let's say in kindergarten, is done because the first grade teacher who will receive her pupils, expects her to "cover" it. This is a type of "education for the afterlife"—always rationalizing today's pedagogy in terms of the next life. (P. 396)

Katz calls this performance because rather than teaching, practice is framed for some audience—the next grade teacher, parents, principal.

Performance leaves students out of the conceptualization of education and positions teachers within a script. It misses the creativity, power, passion, and connection that is at the heart of teaching.

The issues brought together in this section are not new, but the authors illuminate them in a way that reminds us that education, the ultimate in relationship and interaction, can only meet the needs of students if it is mindfully practiced. Focusing the act of teaching in a way that clarifies our intentions, that is aware of unexpected consequences, and that balances attention to the power of the present and the potential of the future might just be mindful practice. And while it will never be easy, it is definitely worth the effort to try.

NOTES

1. These are actual examples. And just for the record, keeping papers for a semester only controls the parents.
2. Although I understand that algebra is now a fair game.

REFERENCES

Bakhtin, M. (1982). *The dialogic imagination*. Austin: University of Texas Press.

———. (1986). *Speech genres and other late essays*. Austin: University of Texas Press.

———. (1993). *Toward a philosophy of the act*. Austin: University of Texas Press.

Dewey, J. (1897). My pedagogic creed. *The School Journal , LIV*(3), 77–80.

Graue, M. E., Kroger, J., & Brown, C. (2003). The gift of time: Enactments of developmental thinking in early childhood education. *Early Childhood Research & Practice*, 5(1), retrieved from http://ecrp.uiuc.edu/v5n1/index.html.

Graue, M. E., Kroeger, J., & Prager, D. (2001). A Bakhtinian analysis of particular home-school relations. *American Education Research Journal*, 3(3), 1–32.

Holquist, M. (2002). *Dialogism* (2nd ed.). London: Routledge.

Katz, L. (1973). Perspective on early childhood education. *The Educational Forum*, 27(4), 393–398.

Katz, L., & Chard, S. C. (2000). *Engaging children's minds: The project approach* (2nd ed.). Stamford, CT: Ablex.

Kristol, W. (December 15, 2004). The defense secretary we have. *The Washington Post*, A33.

Rogoff, B. (2003). *Cultural nature of human development*. New York: Oxford University Press.

PART II

Responsive to What?

Mark D. Vagle

Although Graue (this publication) suggests that developmental responsiveness is more favorable than developmental appropriateness in early childhood education, this section of the book considers the limits of such responsivity in middle grades education. I wholeheartedly agree with Graue that teaching practices should be located in the moment-to-moment contingent relationships between teachers and students. However, I wonder whether linking responsiveness to developmentalism will accomplish this, as the responsiveness seems to be directed toward the young adolescent's development rather than the moment-to-moment contingencies themselves. At first read, developmental responsiveness might appear to be a generous and loving act as the description of this particular developmental stage is marked by hyper-variability among individual young adolescents. However, a more careful reading points to the possibility that the developmental stage is treated as a neutral platform, without consideration of how the stage has been socially constructed. This begs the question: To what ends should middle grades educators' responsiveness be aimed?

I am also a bit concerned about the word "responsive" itself. For one, responsive may imply a linear, binary relationship between adults and young adolescents—first there is an action and then a response or first there is a stimulus and then a response. It is not entirely clear that young people and adults always live this way. At times there is anticipation, guessing, trial and error, mistakes, misunderstandings, failures, successes, and so on. In this way, the term responsive may also be read as hyper-measured and hyper-calculated—as though, for example, the teacher can predetermine where all students are (and should be) at particular moments in time and then simply respond to such predeterminations.

Instead of developmentally responsive, I prefer the phrase *contingently and recursively relational*, as it frees up educators to spend less time seeing young adolescents in developmental frames and more time seeing them in innumerable, lived contexts. I also think that, in practice, it may not matter what a list of developmental characteristics says a boy or a girl should or should not be able to do at a particular time—especially when the list is not implicated as a raced, classed, and gendered document. What does matter is how adults and young adolescents find themselves in relation to one another as they struggle (mightily perhaps) to continually learn and grow with and from one another. For these reasons, I ask those interested in the education of young adolescents to move away from developmental responsiveness in favor of contingent and recursive relationality, as the latter may more accurately capture the intent of educators—to be profoundly present with and for the young adolescent.

The chapters in this section turn our attention toward the particular in that they burrow into specific aspects of middle grades education. They actively resist the developmental age and stage norms and instead try to draw out complexities and nuances that are, perhaps, more difficult to see through developmental perspectives.

In the first chapter, *Pursuing an Answerable Education for Young Adolescents: Implications for Critical Middle Grades Literacy Teacher Education*, I use two theoretical perspectives—Lesko's (2001) contingent and recursive conception of growth and change and Bakhtin's concept of answerability (1922/1993)—to resist the middle school moniker "developmental responsiveness" when determining the kinds of teaching practices one would like to see in middle grades literacy classrooms. In doing so, I create a set of critical, answerable teaching practices for middle grades literacy teachers that draws on critical perspectives forwarded by Moje and Sutherland (2003) and Fecho and Botzakis (2007). These perspectives include the multiplicity of critical discourse communities; raising of questions and the authoring of response by and among all participants; embracing the importance of context and the nonneutrality of language; and encouraging multiple perspectives. I advance these practices by bringing new theoretical perspectives to bear.

In the second chapter, *Black Adolescent Identity, Double-Consciousness, and a Sociohistorically Constructed Adolescence*, Harrison revisits W. E. B. Dubois' classic work, *The Souls of Black Folks* (1903). She argues that the way Dubois captures the complexity of what it meant to be a black person in America one hundred years ago through the image of *double-consciousness* still holds true in black culture today, especially for black adolescents. Harrison is concerned that by focusing more on traditional lines of adolescent development, middle grades education has failed to take up critical perspectives on young adolescence—perspectives that are necessary to ensure academic, social, and emotional success for all students and not just students who represent the dominant culture. In the end, Harrison uses Giroux's *pedagogy of representation* as a methodological tool to analyze two dominant discourses about black adolescents, which leads to her framework of black adolescent identity.

In *Fourteen is the New Thirty: Adolescent Girls, their Bodies, and Sexuality*, Hughes asserts that our modern discourse continues to construct "socially young" teenage girls as asexual, assuming that adolescent girls are not really ready to think and talk about body image and sexuality. By using creative nonfiction (Gutkind, 1997) within multiple genres (Romano, 2000) Hughes draws on her past experiences teaching seventh and eighth grade girls, as well as a girls' forum she created as a safe space for her eighth grade female students to talk about issues of body image and sexuality to further interrogate the idea of developmental responsiveness, readiness, and appropriateness in relation to adolescent girls, sexuality, and their bodies. Hughes opens up spaces in educational contexts so we can begin tapping into the multiple tensions and dilemmas that have surfaced from marginalizing sexuality in relation to young adolescent girls.

The SMART Board as an Adolescent Classroom Technology offers another important perspective on young adolescence and middle schooling in that it demonstrates how adolescence has been and continues to be *usable* when it is understood as an immobile, unified, and homogeneous term used to tell "facts" about adolescents. Drawing on the usability of the SMART board, Bridges-Rhoads welcomes Foucault's discussions (1982) of history, power, and knowledge in order to highlight ways in which linear conceptions of progress and history have been particularly usable in smoothing over the complex and often contradictory understandings and meanings that terms, such as adolescence, might be shown to have if possibilities for difference, plurality, and heterogeneity were allowed free play. In the end, Bridges-Rhoads asks those who work with young adolescents to try to denaturalize everyday conceptions that often seem so unobtrusive that they go unnoticed.

The final chapter in this section, *A Critical Perspective on Human Development: Implications for Adolescence, Classroom Practice, and Middle School Policy*, is by Enora Brown—coeditor (with Saltman, 2005) of *The Critical Middle School Reader*—in response to the previous four chapters. This chapter examines ways in which *critical* and *traditional* understandings of human development and the concept of adolescence inform school policy, and influence teachers' practices and pedagogical "responsiveness" to the learning "readiness" of middle school youth. It begins with a critical perspective on human development, embracing the multiple embedded sociocultural, historical, biological, relational, and psychological processes that constitute human growth and change throughout the lifespan and across generational time. This *critical view* counters the *traditional view* that human development is primarily a naturally unfolding biological process, tangentially influenced by culture. Brown's perspective is based on the assumption that development is mutually constituted by dynamically interwoven *cultural and biological processes*, occurring through interpersonal and societal relationships, communities' historical practices and artifacts, power relations and institutional hierarchies, and meanings ascribed to these processes. Further, this chapter discusses some unexamined assumptions in traditional conceptions— highlighting the need to rethink notions of "developmentalism," utilize critical lenses, enable educators to question normalized developmental views, and envision complex youth constructs and new pedagogical practices in schools.

The hope is that these chapters give those interested in middle grades education pause when thinking about broader conceptions of the education of young adolescents. Although policy is often (and perhaps necessarily even) created based on large overarching ideals, practice needs to be informed more by particular considerations, in particular contexts. These chapters try to be responsive to this commitment.

REFERENCES

Bahktin, M. (1993). *Toward a philosophy of the act* (M. Holquist & V. Liapunov, eds; V. Liapunov, trans.). Austin, TX: University of Texas Press. (Original work published 1922.)

Brown, E., & Saltman, K. (2005). *The critical middle school reader.* New York: Routledge.

Du Bois, W. E. B. (1903). *The souls of black folk.* New York: Bantam Classic.

Fecho, B., & Botzakis, S. (2007). Feasts of becoming: Imagining a literacy classroom based on dialogic beliefs. *Journal of Adolescent & Adult Literacy, 50*(7), 548–558.

Foucault, M. (1982). The subject and power. In H. Dreyfus & P. Rabinow (eds), *Michel Foucault: Beyond structuralism and hermeneutics* (pp. 208–226). Brighton, Sussex: Harvester.

Gutkind, L. (1997). *The art of creative nonfiction: Writing and selling the literature of reality.* New York: John Wiley & Sons.

Lesko, N. (2001). *Act your age: A cultural construction of adolescence.* New York: Routledge/Falmer.

Moje, E. B., & Sutherland, L. M. (2003). The future of middle school literacy education. *English Education, 35*(2), 149–164.

Romano, T. (1995). *Writing with passion: Life stories, multiple genres.* Portsmouth, NH: Heinemann.

CHAPTER 6

Pursuing an Answerable Education for Young Adolescents: Implications for Critical Middle Grades Literacy Teacher Education

Mark D. Vagle

INTRODUCTION

If the education of early adolescents is taken seriously, and if the developmental changes that occur in early adolescence are not simply dismissed as the vagaries of a "bunch of raging hormones" (see Finders, 1998/1999), but are seen as changes in response to changing spaces and relationships that early adolescents experience, then we also will have to reconsider how we teach literacy to young people in middle schools. Concomitantly, the way middle school literacy teacher education is offered also will need to change.

—*Moje & Sutherland, 2003*

Moje and Sutherland's words here prompt any number of important questions for middle grades literacy teachers and teacher educators— questions that reflect my concern that developmental discourses often fall short of realizing their generous and emancipatory promises. How do common sense assumptions about young adolescents' developmental changes obscure the complex, contextually embedded nature of their lived experiences? How might we then interrogate developmentalism enough to allow other perspectives to flourish? And, in turn, how might other discourses such as Nancy Lesko's notion

(2001) of contingent and recursive growth and change and Mikhail Bakhtin's concept of answerability (1922/1993) be used to articulate critical middle grades literacy practices?

In this chapter, I address these questions in the following manner. I begin by first articulating some of the ways developmentalism and developmental responsiveness are used to envision good teaching in middle grades classrooms more generally. Next, I introduce Lesko's (2001) and Bakhtin's (1922/1993) conceptions and then use them to interrogate this vision. This interrogation is not about tearing down common teaching practices advocated in middle grades literacy teaching; rather it is about trying to continually renew such practices. Based on this interrogation, I return to Moje and Sutherland and welcome Fecho and Botzakis (2007) in order create a set of critical, answerable teaching practices for middle grades literacy teachers. I close by reflecting on some of the implications for middle grades literacy teacher education. Although I focus my attention on the teaching of literacy in this chapter, I do think much can be learned from this project in all content areas. However, I felt I needed to start somewhere and that literacy education was the most fertile place to begin.

DEVELOPMENTALLY RESPONSIVE MIDDLE SCHOOLING

The very creation and existence of middle grades education depends on young adolescence being defined as a unique developmental stage in the life cycle. This definition stems from G. Stanley Hall's discussions (1904) of adolescence at the turn of the century and has been reified over the past one hundred years, particularly in the field of psychology (e.g., Flavell, 1963; Piaget, 1952, 1960). Conversations about adolescence both deepened and widened over this time. Throughout the 1900s those interested in the schooling of those in the "younger years" of adolescence advocated for something unique— something, again, responsive. John Lounsbury and Gordon Vars (2003) trace this progression back to 1909 when the first junior high school, Indianola Junior High School in Columbus, Ohio, was created. Lounsbury and Vars continue their historical trace by noting that nine years later, the Commission on the Reorganization of Secondary Education in its report *Cardinal Principles of Secondary Education* recommended the six-three-three organization of schools (i.e., one–six; seven–nine; ten–twelve)—in effect establishing a unique school for young adolescents. By 1945, the six-three-three pattern became the majority practice. In 1963, William Alexander (reprinted in 1995) advocated the term *middle school* in a speech at Cornell University. A decade later National Middle School Association (NMSA) was established and has, according to Lounsbury and Vars, since served as the leading U.S. organization dedicated to the unique needs of young adolescents and now has affiliate organizations in Canada, New Zealand, Australia, and throughout Europe. Over the past thirty-five years, the five-three-four pattern (i.e., one–five; six–eight; nine–twelve) has become the most common school organization structure in the United States.

That said, many middle grades advocates (which I consider myself) are quick to point out that middle schooling is less about school structures and more about the unique ways in which schooling for young adolescents should be enacted. *This We Believe* (National Middle School Association, 2003), a seminal position paper published by the NMSA, sets forth a clear vision for the schooling of young adolescents. At the heart of this vision is the young adolescent and developmental responsiveness. Statements such as "Educators in developmentally responsive middle level schools hold and act upon high expectations for themselves as well as their students" (p. 14) and "Like the young adolescents themselves, the climate of developmentally responsive middle level schools requires constant nurturing" (p. 14) are used throughout the text, reinforcing the primacy of developmentalism.

Young adolescence, at least in the field of middle grades education today, is generally considered to be between ten and fifteen years of age and, ironically, is close to the same age that Hall (1904) used to demarcate his *adolescence* a century ago. It would seem, then, that adulthood has "moved" and adolescence "stretched" over the past one hundred years. What was termed adolescence is now termed "young" adolescence. How does adolescence stretch and adulthood move? One seemingly clear explanation is that society does the constructing and therefore does the stretching and moving. It is through symbolic and material practices over the years. Adolescence and adulthood do not move on their own and then we observe them and name them. Our naming, living, and constructing happen in and over time. We ascribe meaning. This meaning is not neutral and is imbued with power, struggle, success, failure, possibility, tragedy, and the like. To be developmentally responsive leaves a developmental young adolescence untroubled and untheorized. It assumes a single truth or fact—that young adolescence means something "natural" and that we must create middle schooling in response to it, rather than assume that each and every response we make constructs and produces—there is no way it cannot.

Instead, those interested in the education of young adolescents might leave the notion of "developmental responsiveness" in favor of a "contingent and recursively relational" (Lesko, 2001) vision for middle schooling—as the former, reifies a static consideration of growth and change, while the latter prompts a messy, complex consideration of growth and change. This does not mean that I think it is bad practice to be responsive to the needs of young adolescents— quite the contrary, actually. My concern is the locus of the responsiveness. I want those teaching young adolescents and those preparing teachers of young adolescents to not worry much about what a developmental chart or progression says about *young adolescence*, but to *lose sleep* over the contingencies their young adolescents are growing and changing through. I don't want teachers and teacher educators to depend much on developmental discourses to frame their explanations of students and their learning. Rather I want teachers and teacher educators to be profoundly present with their students when *particularizing* such explanations—using the here and now to explain student learning, not what could have, should have, might have been or will be.

CONTINGENT, RECURSIVE GROWTH AND ANSWERABILITY

Lesko's Contingent, Recursive Growth

Lesko's call (2001) for a remade adolescence is based on a rigorous, thoughtful sociohistorical analysis about how adolescence in the United States has come to mean something over the past century. It begins, as most historical accounts of adolescence, with G. Stanley Hall's (e.g., 1904) thoughts in the early 1900s. Hall's views are laced with concerns for protecting the nation, and as such, focus on making sure uncivilized white boys are taught to be civilized white men who can lead the nation. While the women's rights and civil rights movements have presumably changed the political and cultural landscape of the United States over the past century, it is reasonable to concur with Lesko's assertion that our social construction of adolescence has not changed all that much from the days of G. Stanley Hall. Of equal importance is that the middle school movement was founded and continues to launch from the assumption that young adolescence is a distinct developmental stage—an assumption that in effect "freezes" students in time and space without agency, context, politics, or power. Lesko is particularly concerned that a developmental depiction of growth and change, while seemingly intended to be a generous act designed to locate and define what is unique about people at various points in life, actually limits and determines the individual more than it emancipates him/her. Moreover, Lesko believes that by constructing adolescence as a distinct time period, adults in schools and government are allowed to control, study, measure, anticipate, and redirect the individual. Therefore, contrary to common-sense assumptions about the emancipatory aims of middle grades education, constructing and reconstructing a developmental adolescence serves the adults and society more than it serves the adolescents.

Lesko is equally concerned with the absence of context in developmentalism. Characterizing adolescents one way or another void of context reflects Lesko's concern that adolescents are often described "as 'oversocialized,' passive, [and] without critical awareness or active agency" (p. 195). In response to the dominant *development* and *socialization* discourses of adolescence, Lesko offers a discourse that captures, again, the contingent (profoundly contextual and dependent) and recursive (occurring over and over again in and over time) nature of growth and change. This discourse moves the conception of adolescence away from either-or logic and aims to preserve the complexity of living as an adolescent. "Somehow a remade adolescence must take up the contradictions of being simultaneously mature and immature, old and young, traditional and innovative" (p. 196).

Lesko's use of the word "remade" suggests that adolescence is something that has been, is, and will continue to be constructed. This social construction, Saltman (2005) argues, is part of a critical project aimed at a more substantive democratic society, where democracy is "struggled over by different groups with competing material and ideological interests" (p. 19). Taking a critical stance on adolescence allows for such arguments in that it "acknowledges that there are multiple contextually derived perspectives and sources of truth, many of which have been and are subjugated in order to maintain existent relations of power" (Brown, 2005, p. 4). Adolescence should not be treated as a clearly defined stage in life. However, it does not mean there is nothing scientific or biological about adolescence. Instead, a socially constructed adolescence is one that recognizes "that the meaning of biological and psychological realities do become meaningful or relevant in different ways in different social contexts" (Saltman, 2005, p. 16). Using Lesko's notion (2001) of a contingent and recursive growth and change opens up the possibility of seeing *adolescence* in particular ways and therefore also the possibility of seeing individual *adolescents* in these same ways.

Lesko's contingent and recursive conception is important because it simultaneously critiques dominant stage development claims to growth and change and offers a critical alternative that aims to capture what Lesko describes as the blizzard of factors influencing adolescents. I read "contingent" as contextual and dependent, which could, of course, lead to a perpetual state of "it depends" statements. While this could be quite frustrating, I think it is precisely the theoretical and practical move to make. I read "recursive" as something occurring over and over again in and over time. This means that growth and change during adolescence takes place in all of the discrete contingencies one experiences and in the collective contingencies in which individuals find themselves living. These contingencies often cannot be anticipated and they are not necessarily progressive. Adolescents are already, always growing and changing through multiple relations at given moments in time.

Bakhtin's Answerability

Given Lesko's perspective, it seems appropriate to spend some time thinking through how teachers can become well-versed at living in and through contingent, recursive change with their young adolescent students. To me, the first difficult move teachers need to make is to shed the belief that they can rely on a universal determination of what they ought to do in given situations. Shedding this belief requires more than simply saying, "oh, of course good teaching is situational." It means embracing the notion that the "ought-ness" resides and *comes to be* in these situations—and that the ought-ness shifts and changes in often unpredictable ways. This is not to say that teachers do not make particular commitments that they hold fast to, or that consistently run through their pedagogy. Rather, it is to say that these commitments cannot cover everything in and over time. Bakhtin (1922/1993) says that what we ought to do in given situations is determined in our *answerable* acts or deeds.

> Every thought of mine, along with its content, is an act or deed that I perform—my own individually answerable act or deed...It is one of all those acts which make up my whole once-occurrent life as an uninterrupted performing of acts...For my entire life as a whole can be considered a single complex act or deed that I perform: I act, i.e., perform acts, with my whole life, and every particular act and lived-experience is a constituent moment of my life—of the continuous performing of acts...As a performed act, a given thought forms an integral whole: both its content/sense and the fact of its presence in

my actual consciousness—at a particular time and in particular circumstances, i.e., the whole concrete historicalness of its performance—both of these moments (the content/sense moment and the individual-historical moment) are unitary and indivisible in evaluating that thought as my answerable act or deed. (P. 3)

Bakhtin's words here point away from the general and toward the particular—the particular acts and deeds that are performed in and throughout one's life. When Bakhtin says that moments are unitary and indivisible, he indicates that we cannot break down such moments into their constituent parts, nor can we remove ourselves from these moments. We live in and through the wholeness of our acts and deeds—to say we are *answerable* in these acts and deeds is to say that our responsibilities, our ought-ness, our commitments, and our understandings reside in the moments in which we find ourselves living.

On the whole, no theoretical determination and proposition can include within itself the moment of the ought-to-be, nor is this moment derivable from it. There is no aesthetic ought, scientific ought, and—beside them—an ethical ought; there is only that which is aesthetically, theoretically, socially valid, and these validities may be joined by the ought, for which all of them are instrumental. These positings gain their validity within an aesthetic, a scientific, or a sociological unity: the ought gains its validity within the unity of my once-occurent answerable life. (P. 5)

Taking critical perspectives seriously means to continually interrogate what assumptions are made when claiming unitary constructs. My reading of Bakhtin's discussion of unity and answerability prompts a few *critical* ideas. First, Bakhtin is rejecting a universal ought (theory, science, ethics) and is, indeed, proceeding critically. Second, he does not assume a deterministic answerability. In other words, he would never say that there is one way to be answerable, as this would take him back to the very thing he is resisting—universal determinations. Third, although ought-ness resides in our answerability, our answerability does not tell us what we ought to do. Finally, there is answerability running all over the place all the time, regardless of whether we notice, think about, reflect on, or locate it. In this way, we do not *decide* to be answerable in our acts and deeds, we always, already are answerable—we decide to pay attention to our answerability in given moments and then based on our answerability we figure out what we think we ought to do.

Blending Bakhtin with Lesko, one would resist going to a list of young adolescent characteristics or a chart that lays out the proper time for developmentally responsive teaching practices. Instead, one would ask teachers to pay profound attention to their answerability to students. Again, to say that teachers are answerable to their students in the moment-to-moment, everydayness of classrooms is to point at something that already runs through the fabric of the pedagogic relationship. Teachers are answerable to their students, whether they want to be or not. However, living honestly in a state of answerability does not determine what the teacher should do in particular moments. It means that determinations of what is appropriate already resides in these relations—not in overarching policies or proclamations of what is developmentally responsive. To this end, I think it is important for teachers to pay particular attention to how they are answerable to their students through teaching acts, and through this paying attention determine what they think they ought to do—realizing that whatever they notice and decide is necessarily partial, will always fall short, and will always need to be troubled.

Creating a Set of Answerable Teaching Practices

I now use the blended perspectives from Lesko and Bakhtin to create a set of answerable teaching practices. The practices I forward are general and conceptual in nature and, therefore, do not get to the point of being immediately applicable to classrooms—they are ideas for teachers to consider and then, perhaps, enact in their teaching practice. That said, I do provide some examples throughout as a way to illustrate the ideas. These ideas are not original—they draw from some of Moje and Sutherland's (2003) and Fecho and Botzakis's (2007) ideas regarding critical literacy teaching practices. I open each subsection by describing Moje and Sutherland's or Fecho and Botzakis's intentions and then forward how these intentions can also serve as answerable teaching practices in middle grades literacy classrooms.

The Multiplicity of Critical Discourse Communities

Moje and Sutherland (2003) argue that a focus on discourse communities—the notion that texts are constructed and made meaningful in social settings, for social purposes and that the ways one reads and writes depends on such communities—is particularly important "for it is in middle school that young people first experience distinctions among the discourse communities represented by the disciplines or content areas" (p. 151). Further, Moje and Sutherland suggest that young adolescents should not only be able to successfully navigate the literacies in various discourse communities, but also be able to "engage in a 'critical literacy' that can deconstruct, challenge, or disrupt communities, particularly when certain systems or communities operate in unjust or oppressive ways" (Gee, 1997; Luke, 2001).

With regard to the transition young adolescents face when they move from the more "general, self-contained" elementary classrooms to the (usually) more discipline-based middle grades classrooms, Moje and Sutherland (2003) stress that both English/Language Arts (ELA) teachers and teachers across disciplines can plan reading and writing lessons using texts from other content areas and then design critical ways in which student can engage with various texts and produce some of their own. Whether it be reading and writing about an environmental issue in a particular community or conducting a web-based inquiry about communicable diseases, there are opportunities for teachers to make their practices answerable to the students at particular moments in time.

For instance, Moje and Sutherland describe some of the work Moje had done with a middle grades teacher in Detroit. In this example, the teacher had students use multiple literacies to think through the importance of saving the Rain Forest. In doing so, the teacher drew on science and geography content (e.g., air quality issues and regions of the country containing Rain Forests), as well as ELA content and Latino/a culture because many students were of Latino/a heritage. By drawing on these various discourse communities, young adolescents were given the opportunity to move within and among various ways texts are used in contexts. Similarly, Moje and Sutherland suggest that content area teachers from across disciplines can come together to create interdisciplinary units that are important to students—a hallmark of middle grades teaching practices regarding relevance. The *answerability* resides in the depth and degree to which the teacher and student come together around issues that matter— really matter to the young adolescents. Being open to what young adolescents really want to discuss, read, and write about can be a dangerous endeavor. The teacher must decide how "wide the fence" is going to be in her or his classroom. How much troubling and trouble can they handle? Clearly communities and schools have different contextual definitions of when an issue is edgy or inappropriate—just as each teacher does.

Again, the answerability does not reside in a universal determination—there is no best way to be answerable. The same holds true, I think, with invoking critical practices. Critically answerable teaching practices cannot be applied across contexts. Rather, they must be continually interrogated so they live out their very purpose—to be multiple and contextually situated (Brown & Saltman, 2005) rather than single, unified, and decontextualized. This means that the middle grades literacy teacher must adopt a fluid pedagogy that depends on a profound presence with his or her students. At the same time, the teacher must realize that he or she is always limiting, always falling short, and will never achieve an enlightened state.

Dialogic Pedagogic Practices

In Bakhtin's later work (e.g., 1981) he described the dialogic nature of language, and although he no longer uses the term answerability in this later work, I read some of it as having answerability running through it. Fecho and Botzakis (2007) describe five pedagogic practices that are often enacted in dialogic classrooms:

(1) raising of questions and the authoring of response by and among all participants,
(2) embracing the importance of context and the nonneutrality of language,
(3) encouraging multiple perspectives,
(4) flattening of or disturbance within existing hierarchies, and
(5) agreeing that learning is under construction and evolving rather than being reified and static. (P. 550)

Fecho and Botzakis are careful not to offer these practices up as static, "best" practices that must be employed, as that would run against the very Bakhtinian notions they are drawing on, namely heteroglossia, utterance, and carnival. At the same time, they carefully articulate how they use each of these notions to offer up these practices. In this chapter, I briefly sample their first three practices and therefore do not do them justice. I hope, however, that I can forward their ideas by bringing answerability to bear on them as well.

Raising of Questions and Authoring of Response
Fecho and Botzakis (2007) argue that heteroglossia brings attention to context as a factor for meaning and that Bakhtin was particularly concerned with making sure that primacy be given to the

temporal and spatial contexts in which words are used, as opposed to the medium in which the words were, for example, printed. "Meaning is so dependent on context that it remains forever in process, at the intersection of centripetal tensions-those forces that usually represent collective authority and seek to stabilize and center-and centrifugal tensions-those forces that usually represent individual interpretation and seek to diversify and pull outward" (p. 551).

Fecho and Botzakis feel that this view of heteroglossia has significant pedagogic implications for teachers and students in classrooms. For one, it assumes that meaning is constantly under construction through the interactions between teachers and students. This means that the teacher—typically the authoritative voice—and the students—typically the voices of individuals—commit to raising questions and offering responses. In my own teaching of teachers, my students tend to feel that all "good" teaching already has this in place. I remind them, first of all, that we must further examine the *good* in good teaching. Second, I find it necessary to push a bit more on what it might mean to have all players (students and teachers) raising questions and offering responses. I ask what it might mean to be a teacher and allow students to raise questions that are not planned or that might pull one away from his or her lesson plan. If the teacher pulls back quickly or proffers a response that accomplishes the same, then he or she is not "being" heteroglossic. At the same time, I read Bakhtin's heteroglossia much the way I read his notion of answerability—heteroglossia (the tension between the centering and fleeing of language) is already, always happening. Regardless of what a teacher does or does not allow, this heteroglossic tension is at play. The issue at hand, then, is the degree to which teachers recognize these tensions and opens them up for consideration in their pedagogic relationship with students. Fecho and Botzakis (2007) are mindful of this when they say that, "what we feel a working understanding of heteroglossia implies for a classroom is a realization that language and meaning are always in play and that each of us has something to contribute to that intellectual struggle" (p. 551).

Embracing Context and the Nonneutrality of Language
Saying language, meaning, and intellectual struggles are always at play also means that they are never "pure" and "empty," nor are the contexts. By definition contexts are always shifting and changing. Fecho and Botzakis (2007) stress that language is never neutral—it is

always coming from somewhere, always going somewhere, and is unpredictable.

> Educators need to see the empowering aspects of helping students to move from seeing language as belonging to nobody and then belonging to others before they can ultimately claim it as "my word," one they imbued with their expression in contexts from their lives. Learning, if it is to be engaging, must connect meaning to context, must acknowledge the temporal and spatial. What we need to consider now is in what ways and to what extent will conditions prevail that will allow these perspectives to remain in dialogue. (P.552)

Returning to my example of the assumed "good" in good teaching, again, illuminates a significant problem in teaching and teacher education—that everyone knows or should know what good teaching entails as though it exists out there waiting to be uncovered through an archeological dig. Embracing context and nonneutrality of language means that good teaching comes to be in the practice of teaching. Teachers and students are answerable to one another in the pedagogic relation, and this answerability is fully situated in nonneutral contexts. Middle grades literacy teachers must become astute at not only allowing language, meaning, and struggle in their classrooms, but also at making language, meaning, and struggle the very basis of their classrooms. This of course means teachers must turn this nonneutrality on themselves, acknowledging the ways in which their words are imbued with context and belong to others and nobody at the same time. This may involve teachers and students writing together and talking about how their writing is contextualized, how they are using language, how others have used this language, and how they might use that language differently next time and the times after that.

Multiple Perspectives
This progression is not linear, is not moving toward a singular, defined end, and does not belong to the individual—even though the individual may be uttering at any given moment in time.

> Of course, language use is social. Certainly, classrooms are spheres of social activity. However, we wonder to what degree teachers and teacher educators overtly use the social aspects of a classroom to further the learning occurring there...We also argue that the invitation of widespread participation implies the need for multiple perspectives to be in play in the classroom for more than one possible slant to have

efficacy…A result of these links is a sense that knowledge is under construction and always open to scrutiny. One perspective begs the need for other perspectives. One utterance seeks the company of other utterances. One text positions itself within other texts. In our dialogic classroom, learning is seen as cumulative; response builds upon response. (Fecho & Botzakis, 2007, p. 553)

This requires much of the teacher. This means the teacher must become comfortable inviting and actually demanding multiple responses, even (and perhaps especially) when he or she thinks what matters might be settled for the time being. For example, oftentimes when I observe middle grades literacy teachers discuss poignant issues with their students, they are quick to take the first or second response to a question—especially if they feel the response makes good (common) sense. Moreover, when a student utters what is perceived to be an "off-the-wall" response, the teacher quickly dismisses the student. I attribute some of this to the anxieties that often accompany observations and a continual need to cover or get through lessons. At the same time, I think teachers and teacher educators must make a conscious effort to let go of these concerns a bit, so they can allow multiple perspectives to flourish. In my work as a teacher educator I have asked students to write themselves reminders directly into their lesson plans and have asked them to let their class get a little unruly, so that multiple responses can thrive. Inasmuch as tight protocols and structures can elicit multiple responses I find in my own university teaching and in my students' teaching that letting go a bit can also be helpful. Many students won't proffer responses in tightly controlled settings, but might just erupt with something when they are either not supposed to or when no one is watching. Middle grades literacy teachers need to get good at allowing for and noticing such occurrences.

CONCLUSION: IMPLICATIONS FOR CRITICAL (ANSWERABLE) MIDDLE GRADES LITERACY TEACHER EDUCATION

Perhaps some might feel as though Bakhtin's use of the term answerability and my appropriation of it here implies that there is an "answer" to teaching young adolescents, and that pursuit of such a goal reflects a Utopian dream that is not only elusive, but also naïve. To me, Bakhtin is not into answers and would not think that we could ever arrive at one. His goal is to clearly reject a universal application of what one should do in given situations. In other words, no set of laws or rules can govern relations, even though those in such relations might use the rules or laws to make their determinations. The same holds true for those who prepare middle grades literacy teachers. Just as teachers are answerable to their young adolescents, teacher educators are answerable to their students—future teachers. These critical, answerable commitments must be *turned back toward* teacher educators as they are proffered out to the teachers they aim to teach.

Moje and Sutherland (2003) ask middle grades literacy teacher educators to take some of this work head on.

> Middle school literacy teacher education might also engage pre- and inservice teachers in explicit discussions of and practice in recognizing the many different and competing discourse communities in their own lives...so that they might also understand the complexity of those discourses at work in the lives of their students...A focus on the discourse communities of people's lives turns attention toward the discourse communities represented in ethnic, cultural, gendered, sexual orientation, and socioeconomic affiliations, among others. (P.158)

I somewhat resist teacher education *modeling* discourses that seem to reduce teacher learning to a "if you show them how they will in turn be able to go out and do it" phenomenon. This would assume, again, that context does not play a significant role. That said, I think it is good teacher education pedagogy to try to live out the practices one hopes future teachers will employ in their practice. By live out, I mean to take risks with practices, reflect on the practices, make adjustments, and openly dialogue about them with students. Just as each middle grades literacy classroom is a unique, answerable context, so is each university classroom.

When Fecho and Botzakis (2007) wonder whether teachers and teacher educators use the social aspects of the classroom to further the learning, they are saying something quite profound. To say learning is something that is *furthered* implies that learning is something that grows and changes—that it is not something prepackaged to consume. To say that the social aspects of the classroom can be used to accomplish this is to say that the teacher never knows what is in store for the learning when entering the classroom. Granted, the teacher is responsible for getting things started and, in today's policy context, may have scripted starting and ending points relative to state adopted standards and end-of-year tests. At the same time, here is where Bakhtin's earlier work can be utilized as a useful reminder. No matter what outside laws or rules state, the teacher is ultimately answerable to his or her students and their furthered learning. Just because there is a standardized test waiting does not mean the students are not moving in and out of discourse communities that accept and resist such edicts. Teacher educators must help teachers learn to take hold of the standardization so they can play with it a bit, otherwise the standardization will take hold of them—and the answerability to their students will remain obscured.

REFERENCES

Alexander, W. M. (1995). The junior high school: A changing view. *Middle School Journal, 26*(3), 20–24.

Bakhtin, M. (1981). Discourse in the novel. In M. Bakhtin, *The dialogic imagination: Four essays* (M. Holquist, ed.; C. Emerson & M. Holquist, trans.; pp. 259–422). Austin: University of Texas Press.

———. (1993). *Toward a philosophy of the act* (M. Holquist & V. Liapunov, eds; V. Liapunov, trans.). Austin, TX: University of Texas Press. (Original work published 1922.)

Brown, E. R. (2005). Introduction. In Brown & Saltman (eds), *The critical middle school reader* (pp. 1–14). New York: Routledge.

Brown, E. R., & Saltman, K. J. (Eds). (2005). *The critical middle school reader.*

Fecho, B., & Botzakis, S. (2007). Feasts of becoming: Imagining a literacy classroom based on dialogic beliefs. *Journal of Adolescent & Adult Literacy, 50*(7), 548–558.

Finders, M. J. (1998/1999). Raging hormones: Stories of adolescence and implications for teacher preparation. *Journal of Adolescent & Adult Literacy, 42,* 252–263.

Flavell, J. H. (1963). *The developmental psychology of Jean Piaget.* Princeton, NJ: Van Nostrand Reinhold.

Gee, J. P. (1997). Foreword: A discourse approach to language and literacy. In C. Lankshear with J. P. Gee & C. Searle (eds), *Changing literacies* (pp. xiii–xix). Buckingham, UK: Open University Press.

Hall, G. S. (1904). *Adolescence: Its psychology and its relation to physiology, anthropology, sociology, sex, crime, religion, and education.* New York: Appleton & Company.

Lesko, N. (2001). *Act your age: A cultural construction of adolescence.* New York: Routledge/Falmer.

Lounsbury, J. H., & Vars, G. F. (2003). The future of middle level education: Optimistic and pessimistic views. *Middle School Journal, 35*(2), 6–14.

Luke, A. (2001). Foreword. In E. B. Moje & D. G. O'Brien (eds), *Constructions of literacy: Studies of teaching and learning in and out of secondary schools* (pp. ix–xii). Mahwah, NJ: Lawrence Erlbaum Associates.

Moje, E. B., & Sutherland, L. M. (2003). The future of middle school literacy education. *English Education, 35*(2), 149–164.

National Middle School Association. (2003). *This we believe: Successful schools for young adolescents.* Westerville, OH: National Middle School Association.

Piaget, J. (1952). *The origins of intelligence in children.* New York: International University Press.

———. (1960). *The child's conception of the world.* Atlantic Highlands, NJ: Humanities Press.

Saltman, K. J. (2005). The social construction of adolescence. In Brown & Saltman (eds), *The critical middle school reader* (pp. 15–20).

CHAPTER 7

Black Adolescent Identity, Double-Consciousness, and a Sociohistorically Constructed Adolescence

Lisa Harrison

After the Egyptian and Indian, the Greek and Roman, the Teuton and Mongolian, the Negro is a sort of seventh son, born with a veil, and gifted with second-sight in this American world,—a world which yields him no true self-consciousness, but only lets him see himself through the revelation of the other world. It is a peculiar sensation, this double-consciousness, this sense of always looking at one's self through the eyes of others, of measuring one's soul by the tape of a world that looks on in amused contempt and pity. One ever feels his two-ness,—an American, a Negro; two souls, two thoughts, two unreconciled strivings; two warring ideals in one dark body, whose dogged strength alone keeps it from being torn asunder.

—Du Bois, The Souls of Black Folks

In Du Bois's classic work *The Souls of Black Folks* (1903), he captures the complexity of what it means to be a black person in America. Reed (1997) argues that "the 'double-consciousness' or 'two-ness' image has been a remarkable, but variously, evocative characterization of the black American condition for several generations of observers identified with widely different intellectual and political projects" (p. 92). Though written over a century ago Du Bois's concept of double-consciousness has transcended time.

For example, Alridge (2003) discusses the double-consciousness that he encounters as a black scholar within the academy. This two-ness stems from the fact that on one hand, as a black researcher, he feels he has a commitment to uplifting his black community and as a black academic, he also wants to make sure that his work is rigorous and respected within academia. He finds that as a black researcher who does historical research on the education of black people his scholarship has been challenged with concerns of objectivity and presentism—and because of his use of historical research to make connections to current day concerns, voice, and agency. Alridge (2003) asserts that Du Bois's notion of double-consciousness "provides me with a way of situating myself in my research as a member of the Black community and as an academic. It also reframes my position as a methodological tool, rather than a liability, in studying a culture and community that I can understand from within" (p. 32).

Similar to Alridge's struggle, black adolescents are also influenced by this dual state. Although some attention has been given to Du Bois's notion of double-consciousness, a study of this dual identity of black adolescents and its academic implications has, unfortunately, escaped research within middle level education. Focusing more on traditional lines of adolescent development, middle level education has failed to take up critical perspectives on young adolescence (Beane, 2005)—perspectives that are necessary in ensuring academic, social, and emotional success for all students and not just white middle class students who reflect dominant culture. In an attempt to begin to fill this void and push boundaries of traditional middle level perspectives, I use Du Bois' double-consciousness (1903) and Saltman's notion (2005a) of the sociohistorical construction of adolescence to explore black adolescent identity.

Before discussing my framework, I briefly situate my view of identity within a broader discussion of two dominant views of identity: one that has grown out of a developmentalist viewpoint and another that is grounded within cultural studies. I then move on to discuss the influences of Du Bois's and Saltman's work on my thinking with regard to black adolescent identity. Following this discussion, I use Giroux's *pedagogy of representation* (1994) as a methodological tool to analyze two dominant discourses regarding black adolescents and to create a framework of black adolescent identity. I conclude with a discussion of why this framework is necessary and what contributions it holds for the field of middle level education.

SITUATING IDENTITY

There are two dominant views on identity. Some argue that these two views can be broken into essentialist versus non-essentialist or modern versus post-modern views of identity (Bauman, 2008; Grossberg, 2008). The first view of identity is embedded within "Western" thought and conceptualizes identity as an individualistic task of discovering a true, fixed, and consistent core. Grossberg (2008) states that

> the first model assumes that there is some intrinsic and essential content to any identity which is defined by either a common origin or a common structure of experience or both. Struggling against existing constructions of a particular identity takes the form of contesting negative images with positive ones, and of trying to discover the "authentic" and "original" content of the identity. Basically, the struggle over representations of identity here takes the form of offering one fully constituted, separate and distinct identity in place of another. (P. 89)

The second view of identity, which has been largely taken up in cultural studies, is in opposition to the modernist view of it (Holland et al., 1998). In particular, it takes opposition to the notion of a consistent, fully constituted identity. Hall (2008) argues that identities are never unified but rather are fragmented. Furthermore, they are "never singular but multiply constructed across different, often intersecting and antagonistic, discourses, practices and positions" (p. 4).

The framework of black adolescent identity proposed in the second half of this chapter is aligned with this second position on identity. This work builds upon the notion that identities are not stagnant, nor are they embedded in universal and ahistorical qualities—rather they are discursive constructs that are always in flux, constantly transforming and interacting with each other. Though this framework focuses on black identity, it is not based on the premise that black identity is constructed within a vacuum that is exclusive of other identities. To the contrary, "the identity of any term depends, for all practical purposes, totally on its relation to, its difference from, its constitutive other" (Grossberg, 1994, p. 13). With this in mind, my discussion of black identity and the relation to its other will be further explored throughout this chapter.

Although I am broadly situating this work within an anti-essentialist view of identity, as previously stated, this work specifically is influenced by Kenneth Saltman (2005a,b) and W.E.B. Du Bois (1903). The works of Saltman and Du Bois are necessary to make meaning of and to contextualize the multiple identities that are at play in the formation of black adolescent identity. Particularly, Saltman sheds light on the complexity of adolescent identity while Du Bois provides insight into black identity. Both perspectives are needed to make meaning of an anti-essentialist view of black adolescent identity. Saltman (2005a) asserts that adolescence and adolescent identity are both social constructions and to make meaning of them as social constructs it is necessary to take a sociohistorical perspective on how certain "truths" about adolescence came into existence. As Saltman notes "meanings assigned to adolescence are not arbitrary but rather relate to broader material and symbolic power struggles. Social meanings embedded in the construction of adolescence grow out of the particular socio-historical, economic, and political realities from which they emerged" (p. 17).

Although Saltman's (2005a) notion of the social construction of adolescence offers a perspective of how to analyze *adolescent* identity, Du Bois's notion (1903) of double-consciousness offers a lens to explore the construction of *black* identity. Taken together, they provide rich opportunities to examine *black adolescent* identity. Du Bois was a pioneer in reconceptualizing the dominant view of race as a genetic predisposition. Despite the prevailing reliance on biological definitions of race, Du Bois conceptualized race as a sociocultural category (Zuckerman, 2004). His writings emphasized that race was a product of social, historical, and economic conditions rather than a biological trait.

Although Du Bois was a pioneer in providing an alternate view of race, his notion of double-consciousness also has great historical significance; it was one of the oldest analysis of black American identity that was not rooted in racist rhetoric that too often plagued society's perceptions of black people during that time. This dual state is also important for a larger discussion on identity. This concept of twoness "artfully articulates the notion not only that identity is often fractured by numerous social identities and social roles within one being but also that these social identities and roles can sometimes even be at odds with one another" (Zuckerman, 2004, p. 8).

It is important to note that both Saltman and Du Bois call for a critical approach to analyzing the social construction of identity. Saltman (2005a) argues that "this social, historical, and political approach to the study of youth emphasizes that the meaning of youth...is in play and is struggled over by different groups with competing material and ideological interests" (p. 19). To truly understand that adolescence is a social construct one must interrogate the actual construct with certain critical questions such as: Who created this construct? What purpose does this construct hold? When and why has a paradigm shift occurred in regards to the construct? What implications does this construct have for dominant and marginalized groups?

Likewise, for Du Bois (1903), to understand that black identity is formed solely by "always looking at one's self through the eyes of others" (p. 3) is too simplistic and not a means within itself. Du Bois also calls for a critical approach to deconstructing black identity that requires one to ask such questions as: Who is the other? What assumptions does the other have about black people? What are the systematic affects from forming a self-identity based on otherness? What power implications are in play? What hegemonic principles are carried out in one's construction of their black identity? Apple (2004) states that

> Hegemony acts to "saturate" our very consciousness, so that the educational, economic and social world we see and interact with, and the commonsense interpretations we put on it, becomes the world *tout court*, the only world. Hence, hegemony refers not to congeries of meanings that reside at an abstract level somewhere at the "roof of our brains." Rather, it refers to an organized assemblage of meanings and practices, the central, effective and dominant system of meanings, values and actions which are *lived*. It needs to be understood on a different level than "mere opinion" or "manipulation." (P. 4; emphasis in the original)

It is only through these questions and by keeping a sociohistorical context in mind that enables me to analyze and make meaning of the formation of black adolescent identity.

Sociohistorical Context of Adolescence

Arguably, no one person has had as much influence on adolescence as G. Stanley Hall and therefore to many he is considered the father of adolescence. In 1903, Hall published the two-volume groundbreaking book *Adolescence: Its Psychology and Its Relations to Physiology, Anthropology, Sociology, Sex, Crime, Religion and Education*. Out of Hall's publication certain "truths" about the nature of adolescence came into existence that continue to inform current day attitudes about adolescence (Lesko, 1996, 2001; Saltman, 2005a). For Hall, his assumptions about adolescence were based on recapitulation theory. During Hall's time this theory was scientifically based and stated that as a person developed, he/she psychologically reenacted the evolutionary stages from the *human savage animalistic* state to the *modern civilized* man. It was Hall's belief that adolescence was the time in the developmental stage between the primitive state and civilized and therefore much attention was given to adolescence.

Historically and socially situating black identity, it is important to note that Hall's work reflected the racist, sexist, and nationalistic ideologies of that time. Fundamentally, Hall's work was geared toward advancing civilization. For Hall, white adolescent males were the key to this advancement. Hall believed that only the white race (and males within the white race) had the ability to recapitulate into a civilized state. Thus, all nonwhite races were considered "adolescent races" because they could only recapitulate to the equivalent of white adolescents.

Understanding that Hall's philosophy of adolescence was instrumental in maintaining White male superiority (Bederman, 2005; Lesko, 2001) is necessary when one explores black adolescent identity. It is only through an understanding of this context that we can make meaning of current perspectives of black adolescent identity as a social construction. At the center of the constructed adolescence was a notion that all black people only had the capacity to be equivalent to white adolescents. Hall created a paradox where adolescence for white males was a crucial time period between the savage and civilized state that needed to be cultivated and honored because they were the key to the advancement of civilization. Whereas, adolescence for black people—regardless of age—was seen as a stage that was largely defined by their inability to achieve the developmental status of (no higher than) a white adolescent, in essence, it was impossible for black people to even reach a civilized state.

Although today racism is largely seen as a social taboo, one can logically presume that just as Hall created a legacy of how adolescence is viewed in present day society, his legacy of how he viewed adolescent races still has systemic effects on present day societal views of adolescents of color. More so, specifically relating to the construction of black adolescent identity and following a Du Boisian philosophy, Hall's definition of the way that the "other" views adolescent races in effect has created a sociohistorical context in which black adolescents view themselves and how their identity is constructed—always through the eyes of the other. It is within this sociohistorical perspective that I now introduce and then discuss two dominant discourses related to black adolescents.

Introduction of Discourses and Mode of Inquiry

Within society there are several dominant discourses with regard to black adolescents. For black boys some of these include being deviant, academically disengaged, and delinquent (Duncan, 2002; Ferguson, 2000; Roderick, 2003). For black girls some include being promiscuous, loud, overly talkative, and having an attitude (Lei, 2003; Morris, 2007; Tolman, 1996). In order to create my framework I explore two dominant discourses that have been taken up in academia and in popular culture. The first is the "acting white" discourse that gained wide notoriety through Fordham and Ogbu's study (1986) of black high school students in Washington D.C. In this study, Fordham and Ogbu concluded that black students' academic disengagement can be attributed, in large part, to their desire to not be seen as acting white. The other discourse that will be discussed is the notion of the racial self-segregated student that was discussed in Tatum's book *Why are All the Black Kids Sitting Together in the Cafeteria?* (1997).

Giroux's (1994) pedagogy of representation will be used as a mode of inquiry to explore these two dominant discourses of black adolescents. "Pedagogy of representation focuses on demystifying the act and process of representing by revealing how meanings are produced within relations of power and narrate identities through history, social forms, and modes of ethical address that appear objective, universally valid, and consensual" (p. 47). Three questions guided this inquiry: What is the construct of black identity in each discourse? How is black identity constructed in each discourse? How can these constructs be utilized to develop a framework of black adolescent identity? Central to the exploration of these questions is a sociohistorical view that is influenced by, but not limited to, Hall's (1904, 1906) and Saltman's (2005a,b) views of adolescence and are juxtaposed with Du Bois's notion of double-consciousness of black people.

The "Acting White" Black Adolescent

Since the publication of Fordham and Ogbu's study(1986), the *acting white* discourse has received great attention and has been at the center of many debates (e.g., Foley, 2005; Foster, 2004; Gibson, 2005; Ogbu, 2004). One of the major criticisms of the original publication and Ogbu's revisitation of the topic was that not enough attention was given to societal and historical influences in relation to this discourse (Foley, 1991, 2004; Gibson, 2005). However, in defense of their research, Fordham (2008) and Ogbu (2004, 2008) state that their work has been misrepresented over the years. I do not intend to enter into a debate regarding the merits or intentions of their work. Rather, I am focusing on the discourse as something that has been taken up by society—largely in ways that promote cultural deprivation of black youth. Regardless if the acting white discourse is "true" or not, it has been used to represent black youth in a particular manner. Analyzing this construct from a sociohistorical perspective and pedagogy of representation stance, I focus my discussion around the following questions:

1. Which discourse came first?
2. Did the system of inequality within schools lead to a discourse that suggested that being black and academically successful means to act white?
3. Or did the construct of equating the academically successful black student to acting white come first and therefore to maintain an identity as a black adolescent one must choose to be an underachiever?
4. In other words, was the discourse of acting white created because black adolescents were underachieving or did black adolescents choose to underachieve because they did not want to be seen as acting white?

It is important to note that both lines of reasoning are problematic because both create a dynamic where black adolescents equate intelligence with whiteness. Therefore, some black adolescents feel that in order to be smart they must disengage from or compromise their black identity. Although both discourses are problematic, I suggest that the pervasive line of thought that has been taken up implies the latter reasoning—which is detrimental because of the hidden neutrality within the interpretation. It frees society and particularly educators from taking a reflexive look at how they are complicit in perpetuating the "acting white" discourse. To conclude that black adolescents underachieve because they do not want to be seen as acting white places the brunt of academic underachievement on the black adolescents themselves.

From a sociohistorical perspective, education has always been used to uphold certain social and racial orders (Spring, 2007). Historically, education has been strategically controlled for black students because of the fear of its emancipatory powers. This was seen during slavery when it was illegal to teach slaves how to read, during segregation with the allocation of unequal educational resources, and during integration with the tracking of black students out of honors classes and the removal and demotion of black teachers and principals. To borrow a term from Carter G. Woodson (1933/1990), this *mis-education* of black adolescents has created a reality for some black youth—that is, to be academically successful is to be white. Unfortunately the mis-education of black adolescents has often been ignored and the acting white discourse has often been taken up to the detriment of black youth. Thus these socially constructed racial and social orders, largely through the use of education, are often placed onto adolescents and play a crucial role in adolescent identity construction.

It is my belief that the *acting white* discourse is an agency used by some black adolescents to make meaning of their own identity within the operational dialectic between their adolescent identity and their black identity. Black adolescents are not blind to their reality nor are they unaffected by the way they are represented. They can see that there are a disproportionately high number of black students within special education classes and a minute number in honors classes. As Du Bois (1903) stated, "to be Black is to view oneself through the eyes of the other" (p. 3). If black adolescents see themselves and society sees them as black underachievers, then the others in which they view and construct their identity must be white achievers. Therefore, to be black and academically successful means to act white because success has purposefully been created to be synonymous with whiteness. Based on the historical and social context, black adolescent identity is created and recreated. In this case, it is my belief that the acting white discourse of black identity was created partially due to the "mis education" of black students and has been recreated as black students intentionally underachieve because they did not want to be considered acting white.

Although my discussion of acting white centered on which discourse came first, I want to emphasize that the acting white discourse is not as simple as this discussion implies. Apple (2004) warns,

> all too often, we forget the subtlety required to begin to unpack these [the school, the curriculum, and the educator] relations. We situate the institution, the curriculum, and ourselves in an overly deterministic way...It presupposes an idea of conscious manipulation of schooling by a very small number of people with power. While this was and is sometimes the case...the problem is much more complex than that. (P. 3)

To begin to unpack the acting white discourse it is necessary to interrogate the many forces—outside of tracking—that contribute to equating academic success with whiteness. These include, but are not limited to, a curriculum that is often elusive for academically successful black figures, a curriculum that is not culturally relevant to black students' lives, and the constant negative media representation of black youth.

"To be a Black Adolescent is to be Self-Segregated" Discourse

To explore this discourse I will analyze Beverly Tatum's book *Why are all the Black Kids Sitting Together in the Cafeteria: And Other Conversations about Race* (1997) through my three analytic questions posed earlier: What is the construct of black identity in this discourse? How is black identity constructed in this discourse? How can this construct be utilized to develop a framework of black adolescent identity? In relation to these questions it is important to note that Tatum states, "conversely, it could be pointed out that there are many groups of [W]hite students sitting together as well, though people rarely comment about that" (p. 52). In this book, Tatum explores the complexity of racial identity development. In particular, she devotes some of the book to black adolescent identity construction. It is also within this chapter that she provides rich scenarios based on her interviews with students—providing a context to further examine the construction of black adolescent identity.

Tatum reflects on an interview that she had with a student, Jon. Through the course of this conversation Jon moves from being one of the students who did not sit at the black table in high school to a college student who sat at the black table. Tatum argues that "his example illustrates that one's presence at the Black table is often an expression of one's identity development, which evolves over time" (p. 67). This dialogue starts with a discussion of the influence that race had on his athletic choices. Tatum states that his experience is an example of *racelessness*—which is when a person attempts to minimize attributes associated with one's racial identity in order to assimilate into the dominant group. Jon stated

> At no point did I ever think I was [W]hite or did I ever want to be [W]hite...I guess it was one of those things where I tried to de-emphasize the fact that I was Black...I didn't want to do anything that was traditionally Black, like I never played basketball. I ran cross-country...I went for distance running instead of sprints...I quickly

realized that I'm Black, and that's the thing that they're going to see first, no matter how much I try to de-emphasize my Blackness. (P. 64)

Jon's experience shows the complexity of black adolescent identity construction. At this stage in Jon's life he felt the need to stray away from his black identity and focus more on forming an adolescent identity. For Jon it was clear that this adolescent identity was situated in terms of whiteness. It was not that he wanted to be white. Rather, instead of being a black adolescent he wanted to just be identified as an adolescent—something often afforded to his white peers. In order to construct this identity Jon deemed it necessary to de-emphasize his blackness.

It was evident that Jon had a concrete idea of what it meant to be a black student athlete, which is problematic within itself. Jon's idea was defined by fixed parameters of what sports black athletes are typically known to play—largely basketball. In his choice of joining track, Jon realized that to distinguish himself from the stereotypical black athlete he had to choose cross-country over sprinting. However, even in Jon's attempt to use sports selection to mediate between his black identity and adolescent identity he soon realized through life experience that regardless of his efforts he was still going to be viewed as black. His experience shows how identity construction is socially situated. Even in an attempt to construct an identity that would stray away from his blackness and focus more on his adolescence, his social dynamics, did not allow him to construct that identity void of race.

It's like I went through three phases... My first phase was being cool, doing whatever was particularly cool for Black people at the time, and that was like in junior high. Then in high school, you know, I thought being Black was basically all stereotypes, so I tried to avoid all of those things. Now in college, you know, I realize that being Black means a variety of things. (P. 65)

Jon's transition from middle school to high school and finally through college highlights the intricacies and fluidity of black adolescent identity construction. It is in a constant state of flux. For Jon, his identity shifted from solely identifying with what he thought it meant to be black to completely disassociating with his blackness, and then forming a positive black identity, which he came to see as not being limited to stereotypical images of blackness.

When I was in junior high school, I had [W]hite role models. And then when I got into high school, you know, I wasn't sure but I just didn't think having [W]hite role models was a good thing. So I got rid of those. And I basically just, you know, only had my parents for role models. I kind of grew up thinking that we were on the cutting edge. We were doing something radically different than everybody else. And not realizing that there are all kinds of Black people doing the very things that I thought we were the only ones doing... You've got to do the very best you can so that you can continue the great traditions that have already been established. (P. 66)

Jon attributes his ability to construct a positive self-identity to the awareness of black history during college. This history provided strong black role models that allowed him to expand his definition of blackness. He no longer had to accept socially constructed stereotypes of what it meant to be black. Also, he no longer had to look at whiteness as the ideal standard in which to define himself.

Establishing a Framework of Black Adolescent Identity

Based on my analysis of these two dominant discourses I now articulate a framework that can be used to explore black adolescent identity. It is my hope that this framework can be used to provide an alternative perspective about black adolescents that is missing from current research in middle level education. To review, this framework is guided by the following beliefs. First, I argue that Du Bois's notion of double-consciousness of black identity can and should be utilized to explore adolescent identity. It is my belief that black adolescents live in a place of double-consciousness as well: a world torn between adolescent and black identities. Second, I assert that socially and historically situating both adolescence and blackness is necessary to analyze the power dynamics at play, which are often masked as neutral. Adolescent identity has been situated as a white discourse while black identity is often grounded in stereotypical notions of blackness. Finally, it is through a *social construction of adolescence* lens that the complexity of the construction of black adolescent identity can be explored.

Black adolescents construct and make meaning of their own identity and their societal position as black adolescents through constant interactions between adolescent and black identities. This black adolescent identity is a delicate and sometimes not so delicate maneuver situated within a dialectical relationship.

The delicate maneuvers are seen within the hegemonic forces that come into play when black adolescents are trying to make meaning of their adolescence and blackness. They are the systemic effects of such things as institutionalized racism and a Eurocentric curriculum that creates an ideology that whiteness is equivalent to success. These are the unbeknown beliefs that get into the psyche of black adolescents and therefore often affect their decision-making. They are often masked as neutral and therefore usually go without interrogation. On the other hand, the not so delicate maneuvers are the intentional behaviors that black adolescents engage in an attempt to construct a positive black adolescent identity. This was seen in Jon's decision to intentionally join a sports team that was not stereotypical of black athletes.

It is important to note that these interactions are not one-way. For example, adolescent identity interacts and reshapes black adolescent identity. That transformation, regardless how significant it may seem, has an influence on the formation of adolescent and black identity. Adolescent, black adolescent, and black identity are all socially and historically constructed and situated, and therefore each have unique representations and discourses placed on them. Although each identity often stands alone and is represented as an isolated construct, these identities are in continuous dialogue and through each transfer of discourses each identity changes (see figure 7.1).

With regard to this framework on black adolescent identity, I would like to note that many scholars have argued for a perspective that combines race, gender, and class (Bettie, 2003; Collins, 1990; Morris, 2007). Though this framework mainly addresses race and

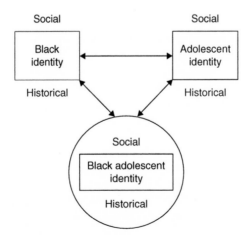

Figure 7.1   Black adolescent identity

age, it is not meant to exclude issues related to, for example, gender or class. In fact, in order to gain a more nuanced understanding of black adolescent identity it is important to see how gender identity and class identity influences black adolescent identity.

EDUCATIONAL IMPORTANCE

Every day, twenty million diverse, rapidly changing 10- to 15-year olds enrolled in our nation's middle level schools are making critical and complex life choices. They are forming attitudes, values, and habits of mind that will largely direct their behavior as adults. They deserve schools that support them fully during this key phase of life. Therefore, National Middle School Association seeks to conceptualize and promote successful middle schools that enhance the healthy growth of young adolescents as lifelong learners, ethical and democratic citizens, and increasingly competent, self-sufficient young people who are optimistic about the future.

—*National Middle School Association, 2003*

To achieve this call, for black young adolescents, middle level educators must take a critical look at the formation of black adolescent identity. Middle level educators must ask critical questions such as: How is or does the middle school design hinder positive identity construction? What role does middle school curriculum play in the hegemonic process of identity construction? To believe that middle schools have no role in the construction of black adolescent identity would be naive. Schools are the very contexts in which much of this construction is occurring.

In a field dominated by white middle class teachers and researchers, to be blind or ignorant to the complex formation of black adolescent identity has detrimental effects on the positive development of social and academic identities of black adolescents. Developing a critical framework to analyze black adolescent identity gives educators an avenue to explore developmental and critical perspectives affecting black adolescents. Also, if adolescent identity operates within a dialectical relationship to black adolescent identity, then by having a deeper understanding of black adolescent identity educators inevitably have a deeper understanding about adolescent identity.

Finally, to understand that black adolescent identity is socially constructed opens up spaces to destroy negative discourses of black adolescent identity and remake positive ones. As Hall (2008) states, "Identities are about questions of using the resources of history, language and culture in the process of becoming rather than being: not 'who we are' or 'where we came from,' so much as what we might become, or we have been represented and how that bears on how we might represent ourselves" (p. 4).

References

Alridge, D. P. (2003). The dilemmas, challenges, and duality of an African-American educational historian. *Educational Researcher, 32*(9), 25–34.

Apple, M. (2004). *Ideology and curriculum.* New York: Routledge/Falmer.

Bauman, Z. (2008). From pilgrim to tourist—or a short history of identity. In S. Hall & P. D. Gay (eds), *Questions of cultural identity* (pp. 18–36). Los Angeles: Sage.

Beane, J. A. (2005). Foreword. In E. R. Brown & K. J. Saltman (eds), *The critical middle school reader* (pp. xi–xv). New York: Routledge.

Bederman, G. (2005). Teaching our sons. In Brown & Saltman (eds), *The critical middle school reader* (pp. 65–86).

Bettie, J. (2003). *Women without class: Girls, race, and identity.* Berkeley, CA: University of California Press.

Collins, P. H. (1990). *Black feminist thought: Knowledge, consciousness, and the politics of empowerment.* Boston: Unwin hyman.

Du Bois, W. E. B. (1903). *The souls of black folk.* New York: Bantam Classic.

Duncan, G. A. (2002). Beyond love: A critical race ethnography of the schooling of adolescent black males. *Equity & Excellence in Education, 35*(2), 131–143.

Ferguson, A. A. (2000). *Bad boys: Public schools in the making of black masculinity.* Ann Arbor, MI: University of Michigan Press.

Foley, D. (2004). Ogbu's theory of academic disengagement: Its evolution and its critics. *Intercultural Education, 15*(4), 385–397.

———. (2005). Elusive prey: John Ogbu and the search for a grand theory of academic disengagement. *International Journal of Qualitative Studies in Education, 18*(5), 643–657.

Foley, D. E. (1991). Reconsidering anthropological explanations of ethnic school failure. *Anthropology and Education Quarterly, 22*(1), 60–86.

Fordham, S. (2008). Beyond capital high: On dual citizenship and the strange career of acting white. *Anthropology & Education Quarterly, 39*(3), 227–246.

Fordham, S., & Ogbu, J. U. (1986). Black students', school success: Coping with the "burden of 'acting white.'" *Urban Review, 18*(3), 176–206.

Foster, K. M. (2004). Coming to terms: A discussion of John Ogbu's cultural-ecological theory of minority academic achievement. *Intercultural Education, 15*(4), 369–384.

Gibson, M. A. (2005). Promoting academic engagement among minority youth: Implications from John Ogbu's shaker heights ethnography. *International Journal of Qualitative Studies in Education, 18*(5), 581–603.

Giroux, H. A. (1994). Living dangerously: Identity politics and the new cultural racism. In H. A. Giroux & P. McLaren (eds), *Between borders: Pedagogy and the politics of cultural studies* (pp. 29–55). New York: Routledge.

Grossberg, L. (1994). Introduction: Bringin' it all back home—pedagogy and cultural studies. In Giroux & McLaren (eds), *Between borders* (pp. 1–25).

———. (2008). Identity and cultural studies-Is that all there is? In Hall & Gay (eds), *Questions of cultural identity* (pp. 87–107).

Hall, G. S. (1904). *Adolescence: Its psychology and its relations to physiology, anthropology, sociology, sex, crime, religion, and education* (Vol. I & II). New York: D. Appletong and company.

———. (1906). *Youth: Its education, regimen, and hygiene.* New York: D. Appletong and Company.

Hall, S. (2008). Introduction: Who needs "identity"? In Hall & Gay (eds), *Questions of cultural identity* (pp. 1–17).

Holland, D., Lachicotte, W., Skinner, D., & Cain, C. (1998). *Identity and agency in cultural worlds.* Cambridge, MA: Harvard University Press.

Lei, J. L. (2003). (Un)Necessary toughness?: Those "loud black girls" and those "quiet asian boys." *Anthropology & Education Quarterly, 34*(2), 158–181.

Lesko, N. (1996). Past, present, and future conceptions of adolescence. *Educational Theory, 46*(4), 453–473.

———. (2001). *Act your age: A cultural construction of adolescence.* New York: Routledge Falmer.

Morris, E. W. (2007). "Ladies" or "loudies"?: Perceptions and experiences of black girls in classrooms. *Youth & Society, 38*(4), 490–515.

National Middle School Association. (2003). *This we believe: Successful schools for young adolescents.* Westerville, OH: National Middle School Association.

Ogbu, J. U. (2004). Collective identity and the burden of acting white in black history, community, and education. *Urban Review: Issues and Ideas in Public Education, 36*(1), 1–35.

———. (2008). The history and status of a theoretical debate. In J. U. Ogbu (ed.), *Minority status, oppositional culture, and schooling* (pp. 3–28). New York: Routledge.

Reed, A. L. (1997). *W. E. B. Du Bois and American political thought: Fabianism and the color line.* New York: Oxford University Press.

Roderick, M. (2003). What's happening to the boys? Early high school experiences and school outcomes among African American male adolescents in Chicago. *Urban Education, 38*(5), 538–607.

Saltman, K. J. (2005a). The social construction of adolescence. In Brown & Saltman (eds), *The critical middle school reader* (pp. 15–20).

———. (2005b). The construction of identity. In Brown & Saltman (eds), *The critical middle school reader* (pp. 237–244).

Spring, J. (2007). *Deculturalization and the struggle for equality*. New York: McGraw Hill.

Tatum, B. D. (1997). *Why are all the Black kids sitting together in the cafeteria: And other conversations about race*. New York: Basic Books.

Tolman, D. L. (1996). Adolescent girls' sexuality: Debunking the myth of the urban girl. In B. J. R. Leadbeater & N. Way (eds), *Urban girls: Resisting stereotypes, creating identities* (pp. 255–271). New York: New York University Press.

Woodson, C. (1933/1990). *The mis-education of the negro*. Trenton, NJ: Africa World Press.

Zuckerman, P. (Ed.). (2004). *The social theory of W.E.B. Du Bois*. Thousand Oaks, CA: Pine Forge Press.

Fourteen is the New Thirty: Adolescent Girls, their Bodies, and Sexuality

Hilary E. Hughes

> *But twilight soon fell upon this bright day, followed by the monoto-nous nights of the Victorian bourgeoisie. Sexuality was carefully confined; it moved into the home. The conjugal family took custody of it and absorbed it into the serious function of reproduction. On the subject of sex, silence became the rule.*
>
> —*Michel Foucault,* The History of Sexuality

Dear authority figures,

Hi! My name is Meg and I am 14 years old. I'm assuming if you're reading this that you're about to teach me or maybe you already have taught me. I was invited to write this opening letter because I believe that what you *think* you know is not always what is going on. (Actually, "invited" has multiple connotations for people, so how about if we go with "strongly encouraged" to help keep it real. My mother recently caught me sneaking out at 3 in the morning to hang out with my boyfriend, so this is one of my adult-enforced-consequences-for-my-actions-kind-of-thing.) Anyway, I'd like you to think about a few things as you read the following (brilliant) chapter written by my language arts teacher.

    First and probably most important—something that really both-ers me about you grown-ups sometimes—is that not all middle schoolers are the same. We come from different backgrounds, have different skin colors, some of our parents are really rich, and some of them are not. We have varying beliefs about God and religion, and some of us have a preference for our own gender rather than being attracted to the opposite gender like we're "supposed to." We *definitely* have different parental-type guardians bossing us around—and yes, some of us even raise ourselves. With everything you already have going on in your teacher/grown-up lives, we feel like you may not be able to see all that sometimes. So this story tries to give you another perspective on what *I* think *you* think you already *thought* you knew. But you might not...yet. Try to be open-minded as you read this, k? Isn't that what *you're* always telling *us* to do, anyway?

So it's basically this: my teacher put together this "safe-space" for any eighth grade girl in our school who wanted to meet because she obviously thought there was a need. (Well, I mean, some of us were into cutting and trying to be all anorexic, so I guess you could call that a "need.") She called it the Girls' Forum (yeah, it was just for girls). We met every other week after school and talked about stuff like guys, cutting, anorexia, bulimia, guys, sex, drinking, drugs, guys, school, our bodies, and other simplistic topics like that. The chapter that follows is her reflection on our forum, which had all different kinds of teenage girls who might not be as *young* as you think we "should be." Ms. H (my teacher, your author) took bits and pieces from her experiences teaching us in the classroom, our time together during the forum meetings, and all the junk she's reading now for her PhD, and she's meshing it all together into one conceptual blob that she's representing through two 14-year-old characters who are kind of all of us from the forum combined. Oh, and she wrote it in different creative nonfiction (Gutkind, 1997) genres because she worships this professor-type who writes books about writing in multiple genres (Romano, 1995, 2000) and because she's just peculiar like that. You know, writing teachers...*nerds*!

I would call the first genre more fiction than creative non-fiction, because, well, none of us liked swimming. Anywho, it sets up this story about the two girls who she uses to represent all of us; other genres (creative non-fiction, I guess) are emails between Ms. H and our parents where she's trying to "explain" us to them; some are emails between her and us; one is a phone conversation between her and my helicopter-parent of a mother (you know, a hoverer); and the more formal ones are journal entries so she can "process," as she says, all the theoretical stuff for her PhD to try and make some kind of sense out of us.

Anyway, just read the thing. Let me know what you think when you're done. I'll put my *Facebook* address on here so you can "friend me," AND so you can see the AMAZING pictures I just added from my spring break trip to Mexico. Well, actually, you might not want to see those. BFN (By for now…c'mon, you knew that one, right?).

Happy Reading! Meg[1]
Soovermiddleschool@facebook.com

* * *

The time of childhood and young adults is meant to be asexual.

—*Lesko, 2001*

Poolside Banter

Meg was sitting on the edge of the pool waiting for Lindsay so they could perform their ritualistic (and somewhat sadistic in Meg's eyes) morning laps. Sitting there, she glanced down at her belly that seemed to be folding into multiple roles in slow motion under her totally cheesy, old-lady bathing suit. Slightly annoyed now, she shifted her thoughts to something even more uneventful—the night before. Meg was thinking about how last night *could* have been; how it had played out in her head, as opposed to how it really happened. He *could* have been like the other guy she had hooked up with—now *that* guy knew how to use his lips. They were warm and juicy as they brushed against her cheek and then slowly found their way to her neck for a while before meandering over to her mouth. Sitting here now, she could still taste his breath because it reminded her of the Grande Skinny Caramel Macchiato she picked up from Starbucks each morning on her way to school. But her experience last night with this kid, who instead reminded her of stale gas station coffee, kept her from reaching the blissful Oh-my-god-state that she couldn't really explain but loved.

God, why did she ditch Caramel Macchiato guy? Her thoughts were interrupted when she glanced down again to see her seemingly elephant-sized thighs below her. They looked gigantic and disgusting just sitting there, so she lifted each leg up alternating them slowly to see if there was any muscle or if it was all just fat. Luckily, her self-deprecating moment was suspended by the sluggish steps of flip-flops dragging on the pavement behind her, and Meg looked up to see

Lindsay. "Hey what's up? Did you weigh yourself?" Meg asked trying to sound supportive because this would be a point of contention.

"Mm-hmm. Yeah. Not exactly what I thought." Lindsay's face had a sheepish grin on it as the words slid out the side of her mouth. She stood next to Meg eyeing the open lanes and the swimmers on either side of them, pretending the conversation was over.

"Yes...*And*...?" Meg asked, sounding a little annoyed, readying herself for a probable blow to her own acceptable body weight.

"Well, I was a little...*off* in my guesstimate yesterday when we were talking." Lindsay answered nonchalantly as she was putting on her cap and goggles.

"C'mon, *fat ass,* how much do you weigh?" Meg barked half kidding and pushing her friend sideways. By no means was Lindsay fat, even overweight for that matter, standing strong in her size zero short-shorts that were rolled down to her pelvic bone and standing only half an inch taller than Meg's lean 5'6 body—a body that had to constantly work on seeing her own size two as acceptable. Meg was not in the mood to play the partial-answers game this morning. "C'mon, seriously, what'd you weigh?"

"That scale was probably wrong." Lindsay replied flatly. "And it doesn't matter anyway, because of this disgusting fat that continues to rub together between my thighs when I walk. It's wretched." Lindsay nodded down toward her thighs as she pulled off her shorts and then quickly jumped into the pool so that her body was not available for public view.

"Are you freaking *kidding* me with that?" Meg smacked water toward Lindsay's face so she'd stop evading her and join the one-sided conversation.

"*Whatever.* I told you, I was a little...off. Maybe five or ten pounds less than I thought. Well, ten, I guess." Lindsay added hesitantly.

"Ten pounds less than you thought?" Meg snapped, gesturing to Lindsay with playful sarcasm in her voice. "I mean, you've been whining like you're a four hundred pound obese lady for weeks: '*I'm soooo fat! I can't fit into* any *of my clothes! I* hate *my body*!' And you come back with *that* half-assed response? You don't like the scale? Ten pounds less than your regular size zero self? You're crazy. I mean it. You're really f'd up in the head!" Meg grabbed a kick board and threw another one to Lindsay so the two of them could continue to chat as they warmed up in the pool and frowned at her friend, concerned.

"I know. *I know*!" Lindsay flinched as she grabbed hold of her kick board and glanced over to her swim partner, acknowledging that she was ready to participate in the conversation as they slowly made their way down the lap pool. "I just thought since I was getting this, this, back fat—I gained like fifteen pounds. I mean look at this!" Lindsay jumped out of the water pushing herself up on her Styrofoam kickboard and grabbed some skin on her waist as if it were a thick slab of pizza dough. Meg rolled her eyes and continued kicking lightly so the two could continue to debate whether or not Lindsay's skin was actually fat or mere skin. It was a debate they had often. "That's called *skin*. It's necessary, I hear, for protecting our bones and muscles."

Lindsay, wanting to change the subject, turned and said, "Speaking of extra skin, didja check out the girl in the other lane? She's got a pound of cottage cheese surrounding her ass and legs. Freakin' disgusting. Swimming obviously isn't working for her. She should try not eating for the rest of her life. So what about you? Have fun last night?"

"Pretty lame." Meg sighed with a condescending smirk on her face. "Stale-gas-station-coffee-guy was about as smart as my 3 year-old cousin. And that kid couldn't turn on a car, much less a girl. Typical of my experiences in life, I suppose." Meg contemplated in a sulky voice, "He was so *lame* when we were making out! I was all, 'Yeah, um, this is fun.' And if he had looked at my face he would have *clearly* seen I was not into it, or him." Lindsay bit her lip to mask her laughter because she knew how important hooking up was for Meg, and if she wasn't comparing the guy to a Starbucks flavor then it was usually a deal breaker.

"AND, by the way, he grabbed my love handles." Meg continued ranting in obvious irritation. "Are you freaking kidding me with that? They should teach them about that in 'How-To-Make-Out-with-Girls-Class or something." Now it was Lindsay who rolled her eyes as they turned for another lap.

"You're so ridiculous, Meg!" Lindsay replied, mocking her swim partner's claim. "Why don't you give the poor guys a chance? If they don't impress you with their brains in the first five seconds are not all over you in the next ten seconds, you tell 'em to go to hell. Not everyone can be as *brilliantly* funny as you or as sexually...*whatever*!" Lindsay quickly glanced over at Meg to see if she was going to pounce on her verbally for calling her out.

"This is *so* very true." Meg agreed with a crooked smile on her face. "But some of them aren't smart at all and I keep them around for awhile...waiting...hoping. Don't I get credit for that? It's the combination of not being smart AND not being able to turn me on! Who wants that combo? I mean, why am I the *only one* in this freaking town who seems to know what I want from a guy?" She asked, waiting for Lindsay to tell her to shut up so they could continue their wretched swim workout. "I'm exaggerating, of course," Meg continued when Lindsay didn't respond, "I couldn't *possibly* know about every single guy's sexual capabilities."

Lindsay almost snorted up the pool water as Meg sarcastically added this last comment. "You come pretty close," Lindsay shot back with jest in her voice. She cut her eyes sideways at her overtly sexual friend, "I'd make a safe bet that if *anyone* could talk about guys' sexual needs, based on a large scale experience, you could. *Slut.*" Lindsay kicked harder in the water to lead them into a faster pace. One corner of Meg's lip curled up into a half-smile while she raised an eyebrow simultaneously. "No need for so many compliments this early in the morning, my good friend. We have all day." She kicked more swiftly to catch up with her lead.

"They're just silly little boys, anyway," Lindsay said as Meg breathed a little harder to try and keep up with the new kicking pace. "I mean, don't we keep hearing that they're developmentally challenged and *years* behind us in maturity?" Lindsay clearly annunciated this last question in her best teacher-mocking-voice, drawing out the word "years" to load it with condescension. "Oh my god, by the way, did you see Elizabeth yesterday? Someone needs to tell that girl that skin-tight cotton Ts aren't for girls with muffin tops. She had lumps and rolls in places I didn't even know existed. Oh, and maybe even more gross than muffin tops, have you seen Shayna lately?" Lindsay asked excitedly, interrupting her own train of thought. "She's like skinnier than...well damn...the day before. Hell, she was skinnier *after* lunch than she was before we ate. I heard she's been going to the gym every morning around five. Then she's running like a ga-billion miles in the afternoons. *Then* she's working out with the high school cross-country team on the weekends. Talk about f'd up in the head. I am sooo not that far gone. I do wonder, though, if she's throwing up or just not eating with all that exercise..."

Meg heard the curiosity in Lindsay's voice and knew that she was fantasizing in an unhealthy way. "Don't even think about it, psycho! You're starting to sound like those crazy anorexic girls that had to go to that loony bin for teens in that book we just read in language arts.[2] Let's focus on getting *me* in shape before you become one of those girls who only eat red Jell-O.

Lindsay noticed the serious but playful tone in Meg's voice and thought she'd join in on the friendly banter. "Well, we can run this afternoon if you're not too busy trying to host a new Sex-Ed for

Dummies class for the entire population of 8th and 9th grade guys in town. *Whore.*" Meg laughed out loud. "Why not squeeze both in before dinner?"

*　*　*

Identities are constructed within, not outside, discourse...pro-duced in specific historical and institutional sites within specific discursive formations and practices...Identities are constructed through, not outside, difference...

—*Hall, cited in Saltman, 2005*

To: teacherfriend@middleschoolassumptions.com
From: concernedmother@adolescentsareproblems.org
Subject: Book

Hey Ms. H,

Meg's mom, Barb, here. I've been thinking about what you said on the phone the other day about adolescent identity *development* versus adolescent identity *construction* and it's starting to make sense. I was wondering if I could borrow one of the books you told me about—*The Critical Middle School Reader*—so I could read more about this idea of social construction. (Where was I in college? I apparently missed all that.) I think the idea of teenagers' cultural and historical influences *is* important because of whom Meg hangs around, because of how her father and I were raised (very differently), the music she listens to, and what she reads. And by the way, that girl won't put down a book to save her life. I think she's read every book on Oprah's list just this month. (Is that OK by the way? Do you think those books are *too old* for her? Should I be concerned?) Anyway, thanks for continuing to talk to me about Meg. I know you're super busy, but if you have time, I'd love to hear more about your thoughts on this adolescent stuff. How did I *ever* get through this *horrible* stage?

Barb

*　*　*

To: concernedmother@adolescentsareproblems.org
From: teacherfriend@middleschoolassumptions.com
Subject: RE: Book

Hi Barb,

I am glad you are interested in reading some more about adolescent identity construction. I think it is important for us as parents and teachers to learn about traditional developmental theories (you know, like we were talking about on the phone with Piaget and Erikson[3]) as well as more recent theories, so we can get a bigger picture of the whole spectrum. For instance, one of the authors in the Brown and Saltman book you asked for wrote,

> To claim that adolescence is a social construction is to call into question more common claims that the nature of adolescence can be understood principally through the sciences of biology and psychology. Instead, adolescence as a social construction suggests that adolescence is a cultural and social category within which facts about youth are interpreted and made meaningful. (Saltman, 2005, p. 15)

I like that explanation of social identity construction because it helps me remember that it is not all about biology and psychology, but also about sociohistorical and cultural influences. I also meant to tell you about the guy who gets the bulk of credit for the initial conception of *adolescence,* because he is far from this "modern" thought we have been discussing. His name was G. Stanley Hall and he was one of those old racist, sexist white boys from back in the day—basically, a sixty-year-old man who invented the thought behind "adolescence" at the turn of the twentieth century by publishing two large volumes on his theories about identity development with 10- to 14-year-olds (Moran, 2000). Hall *only* studied white middle-class boys because, basically, he believed that they were the only ones who were worth it. He decided these middle-class white boys could progress through different phases/stages of development to reach some level of readiness, I guess, that would move our civilization forward.

Interestingly enough (since I know how involved you are with the girls' groups here in the community), Hall alleged that girls and nonwhite teens would never reach the developmental level of the white middle-class boys he was studying (Saltman, 2005). Several theorists in education refer to him as the "father of adolescence," and many of our educational practices in middle school still come from his twentieth century writings (www.nmsa.org). Pretty crazy to think about, huh?

Oh, and this other guy I was just reading that you might be interested in—Jeffrey Moran (2005)—made a good point that really pushed me to think about how we take up an idea so quickly some-times in our culture. Moran wrote about G. Stanley Hall creating that theory called the "recapitulation theory" about adolescence, which was the really twisted theory about hierarchical and racial superiority I mentioned to you before with white boys perched at the top of the food chain and girls/people of color sliding in at the bottom near the Neanderthals. Moran was making the argument that G. Stanley Hall's recapitulation theory became so popular so fast that Americans accepted it as a kind of common sense. It became part of our culture's "mental furniture" as other theories sometimes do in our society with no one questioning them. Great visual, right? OK, have to go grade some papers. Enough theory emailing for now. I'll give Meg that Brown and Saltman book tomorrow.

Cheers! Ms. H

* * *

To: teacherfriend@middleschoolassumptions.com
From: soovermiddleschool@facebook.com
Subject: Uggggg!

Hey Ms. H,

I'm soooo sorry my mother keeps emailing and calling you! She told me I'm supposed to pick up some book for her tomorrow at school. Seriously, can't I send her to Grow-up-and-get-out-of-my-life school or something? I mean, OMG, what a SPACE INVADER! Anyway, I'm lovin' reading *Reviving Ophelia*. It's really good! I can't wait to talk about the body image issues that those girls are having in the book at girls' forum this week. Sooooo reminds me of Lindsay and her crazy body problems!

BFN, Meg

* * *

To say that adolescence is a social and cultural construction is to recognize...that adolescence meant something very different in the past and that it may mean something very different in the future. It is also to recognize that the meanings assigned to adoles-cence are not arbitrary but rather relate to broader material and symbolic power struggles.

—*Saltman, 2005*

THEORETICAL REFLECTION ENTRY 1:
SOCIALLY YOUNG ADOLESCENTS

The things I'm hearing at the girls' forum meetings remind me so much of what I have been reading in Lesko (1996, 2001) about how we tend to marginalize young adolescent girls in our society because of seemingly Victorian influences that still seem to permeate current norms and traditions. Lesko (1996) describes how we continue to keep our adolescents "socially young" because of the age chronologies that we assign them with so much authority. Meaning, the ages we assign them (e.g., Sally is *only 14* so she should not be thinking that way *yet*) keeps them "encapsulated in an age-structuring system that keeps them liminal, or timeless (always 'becoming'), in their adolescent years... [so that they are] both imprisoned in their time (age) and out of time (abstracted), and thereby denied power over decisions or resources" (Lesko, 1996, p. 456).

This historical influence of age chronology seems to attach itself year after year to American cultural norms and further the stereotypes of modern adolescence so that the idea of associating concepts like sexuality with adolescence becomes almost impossible. Taking Lesko's (1996) argument further, it is as if the U.S. tradition has been one of creating (the idea of) socially young adolescents who are not capable of thinking about sexuality and who are not ready to "be" because of a number we assign to them that represents where they *should be* in their thinking, feeling, and doing. Instead, we construct them as young people who will *become* tomorrow rather than acknowledging the many that already *are* today. As our society continues to perpetuate the marginalization of young adolescent girls and their (lack of) interest in sexuality and the body, we might continue to silence the conversations we could be having.

* * *

Silence diminishes opportunities to learn—in school and outside it. Silence withholds questions and opinions that might bring new insights to light. Silence supports the status quo.

—*Sitler, 2008*

To: teacherfriend@middleschoolassumptions.com
From: futurerocknbody@gmail.com
Subject: Girls' Forum Meeting Tomorrow

Hey Ms. H,

I just wanted to tell you that my mom is not really psyched about me going to the girls' forum anymore. I told here we talked about sex the other day and she kind of went crazy! You know she's all religious and has me going to those convocation classes right now, so she doesn't think I should be talking about sex. I really, really, really want to come to the meetings, so would you pllleeeeaaaaassssseeeee email her and talk her into it for me? Just tell her we're not talking about sex anymore or something like that. OK, well, gotta go. Thanks! Lindsay

* * *

To: futurerocknbody@gmail.com
From: teacherfriend@middleschoolassumptions.com
Subject: RE: Girls' Forum Meeting Tomorrow

Hey Lindsay, why don't you have your mom give me a call or email me if she wants to? Tell her my planning is from 9:00 to 10:00. See you later. Ms. H

* * *

THEORETICAL REFLECTION ENTRY 2:
SEXUALITY AS A SILENCED DISCOURSE

What do I say to this mother who refuses to allow her child to attend our girls' forum meetings because the word "sex" was mentioned? It is not my job to argue against parents' decisions about their children, but I feel it is my job to disrupt the silencing of many teenagers' questions. Moran (2000) describes the history of sexuality and sexual self-control in the United States as rooted in a Victorian morality, which descended from centuries of traditional Christian asceticism. It was not just about confining sexuality to the bedroom to silence the practice of sex, but also about being able to define oneself as a respectable Victorian citizen by not speaking of, thinking about, or acting on sexual urges. Even more important than men being seen as respectful citizens within the realm of sexuality, women were believed to be "vessels of purity" by a group of white, male moralist advisers from the nineteenth century. G. Stanley Hall and his white male counterparts claimed that women obviously suffered less than men from sexual urges and even more, "good" women were not *supposed* to have any sexual feelings at all (Moran, 2000).

Born in the mid-1800s and raised by extremely religious parents, G. Stanley Hall was taught as a young child to "refer to his genitals only as 'the dirty place,' and for years he continued to believe that this was their 'proper and adopted designation'" (Moran, 2000, p. 104). Hall's father told him and his young friends about a teenage boy who "abused himself and sinned with lewd women and as a result had a disease that ate his nose away until there were only two flat holes in his face for nostrils and who also became an idiot" (Hall as cited in Moran, 2000, p. 3). As Hall grew older he eventually created a contraption to "prevent his nocturnal erections, and further wrapped his genitals in bandages" (Moran, 2000, p. 3) due to the fear his father instilled in him about the evils of masturbation. Hall continued on in his extremely religious life living piously, but also creating the essence of what we know as adolescence today. I think educational theory that continues to draw heavily on Hall's distorted theory of adolescence is problematic, because it was rooted in distorted religious fallacies and a misconstrued view of racial superiority.

Although the meaning of adolescence can differ according to the culture and historical moments influencing its meaning, G. Stanley Hall's argument was that adolescence was a "crucial period" in which the white middle-class boy's developmental process resembled that of a primitive human species to a civilized white European, and this process was not to involve any kind of sexuality along the way (Saltman, 2005, p. 16). As Moran (2000) suggests, the axis of self-control

> was a range of sexual prescriptions that partook of the biblical tradition of mortification of flesh, but these intensified the self-denial beyond what earlier Americans would have recognized. Sexual intercourse, the advisers insisted, existed solely for procreation; to use it for mere pleasure was supremely selfish and betrayed the continued presence of the brute within the man. Outside of marriage, of course, full chastity was the only proper course of behavior. But even within marriage, partners were not to allow bestial lust to distort their relations. For all Victorians, chastity extended not simply to the body but also to the mind: a lustful imagination was in many ways just as evil as carnal activity. In keeping with the need for mental purity, Victorians decried references to most bodily functions as "vulgar." (Moran, 2000, p.106)

With Hall and his advisers' historical Victorian past continuing to collide with modernity, Bhabha's (1992) disjunctive present seems to parallel how our culture envisions socially young adolescents, and more, young adolescent girls that *should not* be *ready* to talk about topics such as sexuality or their bodies. Maybe I can use Bhabha's idea of disjunctive present when I talk to Lindsay's mother today.

*   *   *

> *A principle of contemporary teaching and teacher education is to begin "where the students are." But where is that? And how do teachers reach the places where students are understood to be?*
>
> —*Lesko, 1996*

To: skepticalmother@churchisgood.com
From: teacherfriend@middleschoolassumptions.com
Subject: Girls' Forum

Dear MaryAnn,

Thank you for the letter and your honesty expressing your concerns about the girls' forum topics. I have to clarify that we did not actually talk about the "act of sex" the other day at the meeting; rather, we were listing topics that the girls would like to discuss over the next few months and *sexuality* was one of the topics many of them listed as a topic. Just in case you are curious, and because everyone is so hip on *Wikipedia* these days (the online encyclopedia that many of my students are using), I checked to see what it says about sexuality in the way I am thinking about it as a topic for the forum discussions:

> Human female sexuality encompasses a broad range of issues, behavior and processes, including female sexual identity and sexual behavior, the physiological, psychological, social, cultural, political, and spiritual or religious aspects of sex. Various aspects and dimensions of female sexuality, as a part of human sexuality, have also been addressed by principles of ethics, morality, and theology. In almost any historical era and culture, the arts, including literary and visual arts, as well as popular culture, present a substantial portion of a given society's views on human sexuality, which also include implicitly or explicitly female sexuality. (http://en.wikipedia.org/wiki/Human_female_sexuality#Feminist_concepts)

Another high interest topic at the meeting was body image in relation to the girls' self-esteem, and Lindsay was actually one of the first girls to bring it up. She said she was fine with me telling you that because she has mentioned her negative perceptions of her body to you before, and she said she hopes to be able to talk to you about it more in the future. I have been reading a lot about body image and self-esteem in relation to young adolescents these past few weeks and the statistics about girls who are not satisfied with how they look are pretty astounding. I was just reading one study that said media in the Western culture has a huge impact on girls' negative body perceptions of themselves, and when they compared 14-year-old girls to boys their same age, the girls had a much higher negative self-image than the boys. The study reported that this was largely due to more media images producing women and girls as extremely thin (Knauss et al., 2007). The girls agreed during the forum meeting, saying that they see so many ads in their magazines, on television, and in the movies that show this absolute obsession our culture seems to have with thinness in women and girls.

Another study I was reading reported that "Early adolescents has been identified as a vulnerable time for girls to develop disordered eating or eating disorders because of the normative challenges associated with that period of development (e.g., physical changes associated with puberty, increased desire for peer acceptance, onset of dating), as well as negative life events in general" (McVey and Davis, 2002, p. 97). I think it is important to talk to the girls about this societal pressure now so they can begin pushing back against those perceptions and feel comfortable with their bodies as they are. If you would like, I can arrange some of the meetings to only focus on body image topics when Lindsay is there, so she can participate in the discussions that will be helpful to her in relation to her body. I will also, of course, respect your wishes if you would like to continue with your decision of not allowing her to attend. Please feel free to contact me if you have any questions.

Sincerely,

Ms. H

*   *   *

The media is only one aspect of an appearance culture that can potentially shape the development of internalization of appearance ideals and body image during adolescence. An important cultural context that has been given less attention in existing research is that afforded by peers.

—*Jones et al., 2004*

To: teacherfriend@middleschoolassumptions.com
From: futurerocknbody@gmail.com
Subject: Good for me, bad for you

Hey Ms. H. Thanks for emailing my mom. I guess you explained girls' forum better than I did so she's cool. Bad news though. Shayna isn't going to come anymore because she says she feels like all the talk about body image is directly aimed toward her and she doesn't want to hear it. I tried to tell her to get over herself—that we're not talking about her—but she won't have it. Not sure what you'll do about that, but you know how skinny she is. She's soooo on the road to anorexia.

TTFN, Lindsay

*   *   *

Growth Opportunities

Meg slumped in the chair, irked as she was waiting impatiently to speak with her teacher. She could hear Ms. H talking to her mother on the phone and she was *so* annoyed that her mother was actually calling to *tell* on her daughter. *So mature*, she thought to herself. Meg couldn't fathom the idea of anyone else's mother calling a teacher to tell on her child. Shouldn't it have been the other way around? Meg wondered how she was even *related* to the woman opposite her teacher's phone. "Oh, she did?" Meg heard Ms. H ask into the phone with a raised voice. "How did you find out? Oh, really? At 3 o'clock in the morning? To meet her boyfriend? Wow, that's pretty late." Ms. H's voice sounded concerned for Meg's mother, but amused for her own sake.

As Ms. H shifted in her seat listening respectfully to the ranting mother, she glanced over at Meg and flashed a warm smile. "Well, it's like we were talking about the other day, Barb, just because you were fourteen once doesn't mean that you know how Meg's fourteen will turn out." Ms. H paused to listen to the retort, "*Of course* I think you're a wonderful mother. If you think about identity construction, the word *construction* itself implies many other social forces are at work, not just your influence, right? So all of the things Meg is experiencing are being created in some kind of social context rather than just unfolding in isolated, developmental sequences. It's not just the decisions you're making, but a lot of other factors as well" (Saltman, 2005, p. 238).

Meg listened to the conversation frustrated. *Of course* how she chose to live her life didn't rely solely on her mother. Meg had friends, she read books and magazines, she watched TV and movies, and most importantly, she loved boys. She also loved having *control* over her body—her curves, her muscles, whether or not she was in the mood to be thin or not—and she also craved the ultimate control she had over the hands that were granted access to explore her lines and curves. Why was this *bad,* she had planned to ask Ms H?

"Yes, I know people call them 'walking hormones,'" she heard her teacher say to her mother, "but like my email said the other day, some people believe it is all about biology, others biology and psychology, and then there are some of us who believe in multiple theories with more influence from outside social forces." Meg could tell that Ms. H was trying to be gentle with her monstrosity of a mother, but she knew it wouldn't help. Her mother didn't understand *anything* about "being" 40, much less 14. Meg looked up at the clock and saw that they only had 30 minutes before school started. Damn her mother. So freaking selfish! Why didn't she just go to a shrink?

Meg looked back over to Ms. H who now sounded like she did when she was teaching a language arts lesson at the end of the day: a little edgy and tired, but still excited to be in a conversation about something that she was passionate. She smiled envisioning her mother on the other end of the phone: pen and paper out, ferociously taking notes so she could quote the lines later to Meg's father when he came home to bring the hammer down. Meg shivered. He wouldn't even look at her before school this morning. She quickly pushed that scene out of her head because it made her feel queasy remembering how mad her father was at her.

"Social mediation. That's right." Ms. H continued with Meg's mother. "That idea that it might not be *all* hormones—or *just* biology—but maybe it's also peer influence, media, you know, whatever cultural influences are contributing to her construction of who she wants to be right now (Brooks-Gunn and Reiter, 2005). What? No, I don't think that's bad."

Meg watched her teacher's expression switch from concerned to impatient and thought about her mother constantly lecturing her through condescension that the feelings Meg had about boys at her age were not "normal," that they were *immoral.* Her mother, who wouldn't let her wear make-up until she went to high school and wouldn't let her get her ears pierced until she was "old enough to date" (how freaking old was that, anyway, *60*?). Her mother, who didn't lose her own virginity until she was married. So old-school, she thought to herself. How was Meg possibly that woman's child? Maybe she wasn't, she quickly thought to herself, maybe she was adopted and they just chose not to tell her.

"Yes, I remember you told me about that book. What was the name of it again? Oh, that's right, *Yardsticks*." Ms. H said casually into the phone. Meg rolled her eyes and sighed loudly wondering if she'd *ever* get to talk to Ms. H this morning. "Well, I know you didn't think the part that said 'sexually active in increasing percentages' under the characteristics of 14-year-olds would ever pertain to Meg, and I think you were right. As Meg has told both of us a few times now, she's not sexually active yet." Whatever that means, Meg thought to herself as she heard Ms. H trying to appease her dinosaur-thinker of a mother.

So what if she was *only* 14. What did being "only 14" mean anyway? She was a hell of a lot smarter than her 20-year-old brother, more mature too, so why did it matter that she enjoyed the feeling of human touch? Meg got up and tapped on her teacher's desk so she could get her attention and pointed to the clock. "You know what, Barb," Ms. H said, obviously interrupting Meg's mother mid-sentence, "I have about 10 minutes before school starts and I need to go get ready. I'm sure you'll do whatever it is that you think is best for Meg's consequences, as well as your own sanity. Email me later and let me know what you think about the Brown and Saltman book. OK, Barb. You're welcome. You have a good day as well. Bye-bye."

Ms. H casually turned to her eighth grade student and smiled, "As you might have some time on your hands in the near future, how about if you help me write an opening letter for a book chapter?"

NOTES

1. This opening letter, along with the genres that follow are (re)constructed based on my continuous reflections of my experiences teaching middle school girls, the girls' forum I created, and my current work on my PhD. Each genre is a representational and theoretical reflection of my experiences with the girls who were a part of the forum.
2. Eliot, E. (2001). *Insatiable: The Compelling Story of Four Teens, Food, and Its Power*
3. See Erikson (1968); Piaget (1952, 1960).

REFERENCES

Bhabha, H. K. (1992). Freedom's basis in the indeterminate. *October*, 61, 46–57.

Brooks-Gunn, J., & Reiter, E. O. (2005). The role of pubertal processes. In E. R. Brown & K. J. Saltman (eds), *The critical middle school reader* (pp. 27–56). New York: Routledge.

Eliot, E. (2001). *Insatiable: The compelling story of four teens, food, and its power*. Florida: HCI Books.

Erikson, E. H. (1968). *Identity, youth and crisis*. New York: W.W. Norton.

Foucault, M. (1990). *The history of sexuality, Vol. 1: An introduction*. New York: Random House (orig. pub. 1978).

Gutkind, L. (1997). *The art of creative nonfiction: writing and selling the literature of reality*. New York: John Wiley & Sons.

Hall, G. S. (2005). From *adolescence*. In Brown & Saltman (eds), *The critical middle school reader* (pp. 21–25).

Jones, D. C., Lee, Y., & Vigfusdottir, T. H. (2004). Body image and the appearance culture among adolescent girls and boys: An examination of friend conversations, peer criticism, appearance magazines, and the internalization of appearance ideals. *Journal of Adolescent Research*, *19*(3), 323–339.

Knauss, C., Paxton, S. J., & Alsaker, F. D. (2007). Relationships amongst body dissatisfaction, internalization of the media body ideal and perceived pressure from media in adolescent girls and boys. *Body Image*, 4, 353–360.

Lesko, N. (1996). Past, present, and future conceptions of adolescence. *Educational Theory*, *46*(4), 453–472.

———. (2001). *Act your age: A cultural construction of adolescence*. New York: Routledge.

McVey, G. L., & Davis, R. (2002). A program to promote positive body image: A 1-year follow-up evaluation. *Journal of Early Adolescence*, *22*(1), 96–108.

Moran, J. P. (2000). *Teaching sex: The shaping of adolescence in the 20th century*. Cambridge, MA: Harvard University Press.

Moran, J. P. (2005). The invention of the sexual adolescent. In Brown & Saltman (eds), *The critical middle school reader* (pp. 103–120).

Piaget, J. (1952). *The origins of intelligence in children*. New York: International University Press.

———. (1960). *The child's conception of the world*. Atlantic Highlands, NJ: Humanities Press.

Romano, T. (1995). *Writing with passion: life stories, multiple genres*. Portsmouth, NH: Heinemann.

———. (2000). *Blending genre, altering style: Writing multigenre papers*. Portsmouth, NH: Heinemann.

Saltman, K. (2005). The construction of identity. In Brown & Saltman (eds), *The critical middle school reader* (pp. 237–243).

Sitler, H.C. (2008) Writing like a good girl. *English Journal*, *97*(3), 46–51.

Wikipedia. (n.d.). Human female sexuality. Retrieved May 5, 2008, from http://en.wikipedia.org/wiki/Human_female_sexuality#Feminist_concepts.

Wood, C. (2007). *Yardsticks: Children in the classroom ages 4–14* (3rd ed). Turner Falls, MA: Northeast Foundation for Children, Inc.

CHAPTER 9

The SMART Board as an Adolescent Classroom Technology

Sarah Bridges-Rhoads

If you ever find yourself using our products and not noticing any-thing, then we have done our job. This is usability at its best. The product is unobtrusive and serving your needs

—*Nancy Knowlton (2006a), CEO of SMART Technologies, Inc., describing her product, the SMART Board, the leading brand of Interactive Whiteboards.*

Please grant me a moment to tell you a tale of usability at its best.

"You don't know what a SMART Board is?" said the student teacher whose lesson I was to observe in just a few moments. Her words, spoken with the delicacy of one who did not want to risk angering her supervisor right before an observation, reflected only a slight hint of disbelief. Her face, on the other hand, asked the question I suspected she really wanted me to answer—*When was the last time you were actually in a classroom, moron?*

Hesitating just momentarily, I threw her back an answer, shaking my head as I spoke. "I have never used or seen a Smart Board." I attempted to mask my words with a feigned confidence I had hoped would hide the internal fretting that had just begun as I wondered whether or not I was, perhaps, a bit behind the times—so-to-speak—having not been teaching for a few years and having just begun this job of supervision. My student, though, needed no words to indicate that I had not succeeded in convincing her that I was even remotely technologically savvy. Her smile, a gentle smirk of sorts or a pout of pity, let me know instantly that I was not up to her standards of excellence as a supervisor.

"I was no moron though," I told myself as I entered the fifth grade classroom behind the student teacher and took a seat in a back corner just moments before the SMART Board lesson began. Did it *really* matter that I had never heard of this newest technological tool that my student had just assured me on our way into the classroom was the "coolest in the world" and was perfect for engaging her young adolescent students who often had other things on their minds besides school? Just in case it did matter, I took a quick glance around the room—hoping to see whatever was so amazing and innovative that would help me progress to the new stage of smartness that my student teacher and her students apparently had already achieved.

Same old, same old. The desks in neat rows facing the front resembled those you might see depicted in *Little House on the Prairie*, except that chatty young adolescents replaced the mixed aged bunch from the one-room schoolhouse, and at the head of the classroom hung something other than that old dusty chalkboard used back-in-the-day—a pristine white board I assumed was the SMART one. I stared at it, not wanting to look away for fear of missing something important. I soon realized though that even if I had turned away I wouldn't have missed much except a lesson that seemed just as old school as that chalkboard on the *Prairie*. As my attention turned back to my student's lesson, I saw she was conducting the familiar come-to-the-board-one-at-a-time-and-wait-to-be-corrected-by-the-teacher dance, while the SMART Board (which turned out to be a computer image projected onto the whiteboard and acted as a touch sensitive control panel for the computer) seemed to be doing nothing but getting in the way.

Each time a child would attempt to "write" a problem on the board with the special SMART pen that seemed to mesmerize everyone in the room as it simulated ink on the board, his or her body would accidentally block the stream of light emerging from the hanging projector. This prevented the child and the rest of the class from seeing anything but a whiteboard. This order of events sent the class into torrents of laughter that would not abate until the child at the board successfully contorted his or her body in any number of ways to avoid the stream of light or my student sent out a warning phrase like "You are in the fifth grade. You are more mature than *that!*" As I watched one child simultaneously try to squat down below the board, balance on one leg, and reach his arm to the height of the problem, I caught myself giving my student the same pouty smirk she'd given me earlier.

Did my modern-day, technologically smart student not notice how this SMART Board was not exactly working?

It is precisely this sort of question that I imagine Nancy Knowlton, the CEO of SMART Technologies, Inc., hopes to eliminate in her quest to make the SMART Board a prime example of usability at its best. In her one-page article entitled *Interactive Whiteboard Usability*, Knowlton (2006a) describes this usability as the "absence of little annoyances" with the way products work—annoyances, I presume, like the one where children's bodies block what is actually being projected onto the board. Ironing out these little kinks helps make a product so uncomplicated that its usage seems intuitive; delivering what we "want and need" at such a level that we don't even have to *notice* what we are using. It is like the weather, she says. When it's "perfect...we don't sit back and think 'I am neither hot nor cold.'" Instead, "we simply enjoy it."

As I sat back at my computer—not exactly enjoying myself the evening after the observation I described earlier—I happened upon Knowlton's (2006a) usability article on the SMART Technologies webpage. I was in the midst of my pursuit to become more knowledgeable about the newest technologies in schools. And it really annoyed me. Her methods of "remov(ing) usability hassles" by observing how people use the product, asking them questions about their frustrations, and working to make everything no more than just "two clicks away" struck me as condescending in a way—as if teachers were not smart enough to problem solve or decipher the various contingencies that come their way, and moreover that teachers needed others to smooth things over. It reminded me of watching *Star Wars* and having someone playing Jedi mind tricks on us by waving their hand and saying "you will not notice this board...it already meets your needs" [some of which were dissemination of information (Knowlton, 2006a,b), global competitiveness (Knowlton, 2006c, 2007), and engagement and management of students of various ages (Knowlton, 2008, 2009)].

After reading it a few times through, I began to wonder how my interaction with my student teacher might have gone differently if Knowlton and her team of collaborators had already worked out the kinks with the SMART Board and made it some force of ease that went unnoticed during the lesson, meeting the needs of the users. Would it have altered, for example, what I considered to be a disastrous post-observation conference in which my student's extreme excitement about the SMART Board repeatedly thwarted my attempts to encourage her to think of more creative and engaging ways to interact with her students and with the board? Would the absence of annoyances with the way the board worked really have made the whole experience one of ease and enjoyment—one where we did not even notice the board or how it worked? Furthermore if we aren't noticing this one aspect of classroom life, what else are we not noticing?

It is these sorts of questions about what we do and do not notice that I examine in this chapter. I follow the lead of critical educators such as Enora Brown and Kenneth Saltman (2005) who are already questioning these processes of trying to see what is not being noticed in various areas of education and society, such as in middle grades education, and who find this type of work necessary for opening spaces for reimagining schooling in support of equitable relations. Drawing on similarities between developmental progressions of the board and adolescence, I expand upon Knowlton's conception of usability and her focus on smoothing out the characteristics of the technology and making the relationship between the user and the technology one of ease.

More specifically, I look to what is being said, thought, and acted upon *in relation to* the board, its supposed characteristics, and the problems associated with it. Returning back to the vignette and other examples from popular culture, I draw upon Foucault (1978, 1979, 1982) to highlight how a technology of power such as unilinear development through time makes the board, as well as adolescence, both usable concepts for disciplining and managing the self and others. I do this because technologies of power cannot be isolated in one community or institution but instead permeate society—and in this situation, permeate the vignette in multiple ways. In this process I highlight how the assumed linear progression of development creates *a sense of being behind* that can be useful in telling "facts" about a person by offering a "desired" direction to head toward and an "at a glance" vision of where one stands. I end with a discussion of how a focus on the smoothing over and ironing out the kinks in the characteristics of ideas or products limits our abilities to imagine schooling that will support individual children or teachers (in the moment) as opposed to worrying as to whether or not a given teacher or child is behind.

USABILITY EXPANDED

I begin this process by returning to Knowlton's (2006a) conception of usability and my hunch that in this vignette she might think that usability has not yet arrived. Since her usability focuses specifically on the relationship between the user and the product—with an interest in the moments when the board was physically being used—there still seems to be a need for work to be done in the process of meeting needs seamlessly. More work is also necessary in order to ensure that the technology itself can be counted on to work efficiently in multiple contexts as a solution to any number of problems its users have previously expressed a need to remedy. If I sent Knowlton a copy of the vignette perhaps she would fast-forward to the moments where my student teacher and her fifth grade students were actually interacting with the board—seeking out the useful information that would help iron out those kinks in the characteristics of the board. That way, it could become a little less complicated, a tad more appropriate for the needs of the lesson, and hopefully *not* necessary to notice (or speak of) anymore.

But, if we pause this process of fast-forwarding to those places where certain objects or concepts are being worked on externally, and instead look to those moments when the board or other objects and ideas are actually being put into discourse (accepted ways of speaking, thinking, and acting in certain contexts), a whole new world of usability opens up. It is here that spaces become available for noticing not only which types of knowledge are being promoted and maintained as good, true, or normal, but also for noticing how certain types of knowers (or subjects) are being produced at the same time. We can then see and analyze how the board and other technologies are usable in controlling and disciplining our relationships with ourselves and others as it works on other "problems" that have gone unnoticed—specifically the problem of feeling behind.

Foucault's analytics of power (1978, 1982) is helpful in such a process of noticing. For instance, he analyzes power not only as a repressive or negative force that can act on something or someone from above or below (like Knowlton and her company working tirelessly to tweak a technology to perfection), but also as a productive force that is dispersed throughout the social body and distributed and maintained through various techniques or what he calls *technologies of power* (Foucault, 1978, 1982). It is through these technologies that subjects, that is, human beings, learn to administer themselves and represent themselves as knowers of a certain kind. This in turn allows for the management of groups of people and relationships with selves. Thus, instead of focusing on efforts to make concepts more natural, Foucault's work (1978, 1979) looks to what is spoken about and already being said, thought, and acted upon in relation to certain concepts (e.g., the SMART Board itself is a solution to our problems).

Speaking of the Board with Foucault

My student's question—"You don't know what a SMART Board is?"—reflected her shock at my lack of knowledge of something that I assume she thought the typical supervisor *should* know immediately. This threw me into a panic. Recall how I began to wonder about my status as a moron, almost intuitively it seemed, without needing anyone to tell me to do so. Despite the fact that I had no clue as to how or if the board actually worked, I still *felt* inadequate. I still felt *behind*. Any amount of work Knowlton had done to perfect the ease with which a person actually used a board when standing in front of it could not be blamed for any shortcoming I was feeling. There was something else at work here. Some other unnoticed usability already in play, which was not about my relationship with the board (how well it worked, etc.) but was somehow related to my relationship with myself and with my student.

Power here is productive, not some external repressive force that works on something to change, tweak, or smooth over. My focus at this moment in the vignette and during those subsequent moments when I caught myself staring at the board, going home to research the board, or asking anyone I saw in the next few weeks what they knew about the board did not center on the characteristics of the board itself or on knowing how well it worked in certain contexts. Instead, I was interested in finding information that would make me stop feeling behind. I wanted to become someone who knew about the board regardless of whether or not I would ever touch one. That way I would feel normal, typical, and up to the standards of a technologically savvy supervisor. What is being produced is not only knowledge about the board as I instigate conversations and talk widely about it, but also a specific way in which I can understand myself as one who is behind as a knower, and one who is in need of a certain type of knowledge in order to be deemed normal.

For Foucault (1982) such needs can always be questioned, for he sees the self as not the self of humanism that has an essential nature with natural needs, but instead as subjects whose possibilities for being are made within power relations. Foucault advises us to look to these minute everyday practices (like this moment of self-checking in the vignette) within which power is working upon our actions to shape our conduct (and that of others) and our very sense of what we seem to naturally need. In regards to adolescence, for instance, Nancy Lesko (2001) turns to Foucault to help her ask questions such as,: "What are the systems of ideas that make possible the adolescence we think, feel, and act upon?" (p. 9). Attempting to locate the channels power takes (these technologies of power) to enter into discourse and reach our "most tenuous and individual modes of behavior" (Foucault, 1978, p. 11) becomes important work. With power as productive and running through the social body, these technologies cannot be isolated in one community or institution (Dreyfus & Rabinow, 1982) and instead act as an ensemble of practices that shape and guide our way of being human (Rabinow & Rose, 2003). In relation to my project, Foucault would not look to an individual *piece of technology* like the SMART Board or to an individual concept that seems so stable like adolescence, but to the technologies of power to see how they put certain ideas into discourse.

Speaking of a Developmental Adolescence and the SMART Board with Foucault

For the remainder of the chapter, I focus specifically on a technology of power that permeates the introductory vignette—that of development through time. In linear developmental discourses a sense of always progressing toward something or someplace (such as toward global competitiveness or toward adulthood) can leave people behind or ahead of stages. This makes one able to say something about oneself—I am either behind or ahead of where I should be. This technology of development permeates other discourses as well, making education in particular difficult to talk about in any other way (i.e., No Child Left *Behind*). This developmentalism works through multiple relations of power that help create norms of speaking where one can, for instance, say that a SMART Board is progressing toward smartness as if it were an adolescent progressing toward his/her smartness—adulthood. For this reason, it is helpful to look at how technological development relates to adolescent development—in that similar work is done on those involved.

With regard to developmentalism as a technology of power, there is a sense of development in stages hinging on conceptions of time and progress; this in turn tells us that development marches on in a predictable manner known to have *solved problems* throughout time. To illustrate this point I first return to the SMART Board. The board as a technology has progressed through its own unique stages of development from a slate, to a chalkboard, to a whiteboard, and now to a SMART Board. What comes next, we do not know. But along the way there is a sense that many problems have been solved and will continue to be solved as the board continues to develop through time. The first handheld slates were created to help teachers dole out different problems and assignments to a variety of children in a one-room schoolhouse where children of multiple ages worked at one time. At a time when paper was expensive, the slates were a cheaper alternative (Slate, 2009). Next, with the move to aged classes the invention of the chalkboard—created for military purposes for the easy dissemination of information—became an even better tool for teaching efficiently. There were problems with the dustiness of the chalk getting into the lungs of its user. So the next big improvement was the dry-erase board, with its dry erase markers. No dust chalk while erasing, but still not much more interactive than the chalkboard (Ergo in Demand, 2000–2009). The advent of the Interactive Whiteboards in the 1990s solved this problem. Whiteboards were particularly useful for companies to design creative presentations with the goal of making the dissemination of information more timely and efficient (Knowlton, 2007). Clearly, usability and solving problems are at the center of this development—a linear progression moving closer and closer to perfection.

Like the SMART Board, young adolescence has the same sort of status as a unique stage in life (time). Age is what seems to matter in adolescence, making the ages between ten and fifteen a unique time that Lesko (2001) would say is often said to have its own "demands, problems, meanings, and crises" (p. 107). When seen in relation to other stages—childhood and adulthood, for instance—adolescence becomes something that seems naturally more mature than childhood and less so than adulthood. Certain biological changes mark this stage as something notable, taking with it assumptions that emotional and moral developments have a stage of uniqueness here as well (Saltman, 2005). These simple stages allow for familiar statements such as—"Oh he must be at *that* age" or "When he is an adult…he will know." These statements are used to describe how a youth has outgrown or has yet to outgrow the problems of childhood or adulthood (such as immaturity).

At a Glance

With this sense that development occurs in stages that are unique, there is the idea that the board or age itself is something that can be used to tell facts about a certain school and person. Foucault (1982) uses the term "dividing practices" to describe the type of work that technologies such as developmentalism do to organize, classify, and structure potential ways of being. In other words, we do not necessarily need a top/down mechanism to tell us to use something or not—we just know. These practices work with other mechanisms of power such as visibility to discipline subjects into certain types of people (Foucault, 1979). *At a glance*, for instance, one can tell if a school, a classroom, a teacher, or oneself is smarter based on their boards. It is important to recognize that this does not require any one person to put this on us and force us to think in a certain way.

As I explained earlier, when I asked around to see how others were speaking about the board, I began to notice what seemed similar to what Foucault (1979) has called a "discursive explosion" where, among other things, there was "an institutional incitement to speak about (some idea)" and "a determination on the part of the agencies of power to hear it spoken about and to cause *it* to speak through explicit articulation and endlessly accumulated detail" (p. 18; emphasis added). It felt like everywhere I went people seemed to have stories about how much smarter a school automatically was if the board was hanging on the wall. I heard a teacher candidate, for instance, describe a place she wanted to work in but was apprehensive because it only had chalkboards. I heard stories of children who wanted specific (good) teachers because they had SMART Boards in their rooms. I heard teachers complain that their district was so far behind without them. One parent group in the area even held a five hundred dollar/ person dinner party for parents in order to fund their "Get Smart" campaign, which would help purchase the expensive boards for each classroom. The board itself seemed to be saying that the school or teacher was good.

This *at a glance* knowing also works on adolescence in multiple ways—again in a linear fashion in relation to other stages. This is most evident in discussion of the uniqueness of puberty and its relation to social behavior. It becomes useful to tell facts (i.e., if adolescents are more or less mature, too fast or too far behind) about adolescents such as in the vignette when my student told her students that they were in fifth grade and thus they were more "mature" than how they were acting. It also becomes useful when she beckons aged information to explain to me that youth often had other things on their minds at this age (which I can assume to be the opposite sex due to the talk of raging hormones at that age). Unlike the SMART Board though, where it might not be important for a school to move through the same stages that the board had passed through (i.e., a school jumps from chalkboards to SMART Boards without anyone thinking twice about it), with adolescence the stage itself must be passed through. There is an added sense of the importance of development through time within the stage of adolescence itself—making it quite usable for checking the progress of self and other as well.

For instance, in the movie *13 going on 30*, Jennifer Garner's character is miraculously transported from the body of her thirteen-year-old self to that of a thirty-year-old where she wakes up in the bed of her thirty-year-old boyfriend. Her mind remains thirteen though and the shock of the movie is what would happen if you suddenly were forced to act more mature and more developed. By skipping the majority of adolescence—the span of years that would have allowed her to progress from child to adult—Garner's character runs the risk of not being ready for the place in which she now finds herself. She missed the *unique* stage of development that would have prepared her for what she now faces in the *unique* stage of young adulthood. She has a problem of being either too mature when she returns to her old body or too young in her new.

Additionally, this developmental technology of power is productive in that it tells facts about ourselves and regulates our feelings of normality or abnormality. Although I am confident that one morning back in grade school my twelve-year-old body woke up seven-and-a-half inches taller than it had been the day before, I most likely developed in a step-by-step progression toward the grand adulthood. I, like most of my peers, developed in and through *time* at a specified rate—not too early, not too late, just on time. That said, a problem is produced when this rubric is used to describe one's progression. For instance, it is challenging to deal with the frustrations of not feeling on time. This frustration in not knowing when something will arrive—like a train at the station—is how I felt when waiting for those breasts to finally arrive on my chest before I reached high school or college. I wanted to feel like I was moving in the right direction and on time—like that train was supposed to be.

Within these stages, so much is already neatly timed and you know when and how arrival will happen. If one is patient enough, they will most likely develop through the stage (whereas with a SMART Board, the money a school district has is a factor in progression). As a young adolescent, I used to obsess about the boys who bumped into my little nubs that I wished were breasts, giggling an "Oh that's you, Sarah, I thought that was the wall." I would then run frantically into the bathroom to stuff tissue in my training bras. When my teachers caught me with little tufts hanging out my sleeve, in the name of adolescence, they'd tell me to "calm down," to "give it time." I'd develop, they'd say. I'd make that adequate yearly progress if I'd just stop worrying about having my "wall" a little bumpier after all the other girls. *Adolescence* already knew when I'd get there—it had been charted.

Charts are made for the timeliness of pubic hair, for when girls should begin menstruating and are widely available to adolescents during sex education courses or in the chapter on reproductive organs in their textbooks. This charting has been going on for some time, especially in the fields of biology and psychology and has been useful for helping determine who was on track. They become something to check one's students against and they can check themselves against it as well. If one does not fit like those in the chart, it might seem natural to feel behind and ask oneself any number of questions about one's development. Am I being a good adolescent today? Am I on track? Being on track seemed better, and this was *known* at a glance because *one look* gives the information that takes years for youth to experience. Such immediately consumable facts make adolescence able to be commodified, represented in images and sound bites that can be immediately understood.

Panoptical Time

With her discussion of panoptical time, Lesko (2001) provides a useful visual of the complex processes through which development through time works to maintain and produce certain types of selves. Like the panoptical shaped prison that Foucault (1979) came across in his research on the history of the problem of punishment in French society, one does not need someone standing in a watchtower to force you to do something. We do it to ourselves. This panoptical prison—named for its shape—had a tall guard watchtower, centrally placed in a large space with the cells of the prisoners surrounding it on each side. A guard could sit in the tower and see everything *at a glance,* kind of like a teacher who stands at the corner of the intersection of multiple hallways so she can easily glance down each and see who is up to no good. Or picture that great hall in the Harry Potter movies where all of the professors sat on that raised platform where they could see all the tables without straining too hard. The genius of the design appeared to be in this ability to see all in a quick glance and have them see the guard, knowing that they couldn't escape. If they did, they would be seen and shot down like that young lad in *Shawshank Redemption* who was spotted trying to escape halfway up the fence by the convenient light attached to the top of the tower.

Although Foucault (1979) agreed that this issue of being able to watch and monitor prisoners was important in disciplining inmates, he took it in a different direction. He said that the "procedures of power" that were at work with this panopticon were "much more numerous, diverse and rich" (p. 148). Just knowing that the tower was there was enough to make the prisoners discipline and check themselves. Each person became his or her own overseer. It's like a cop car. You're cruising along the highway and just the mere sight of it makes you slam down on the brakes. But then, you get up closer and the cop isn't even in his car. He's unseen. And sometimes, there are certain parts of a routine drive for which we don't even need a cop to be present in order to prompt our slowing down. We have seen so many cops in that one five-mile strip that we don't even need the visual. We interiorize their gaze.

According to Lesko (2001), time is like that in adolescence as well. It is like that cop car or that panopticon prison that works to structure our experiences and make us do and be certain things without anyone even sitting in it. With age as a great template to judge the appropriateness of actions and like the number on a watch to be passed through each hour, adolescents can always be moving toward the next stage, toward "adulthood," always waiting to arrive. There is no need for anyone to be watching because the idea that development always occurs in and through time in a linear fashion is already so pervasive that we can use it to check ourselves and others, ensuring that while you sit waiting for your SMART Board, you will always feel behind.

CLOSING THOUGHTS

Throughout this chapter, I have used one vignette—one snippet of an experience—to help illustrate ways in which what we do *not* notice (such as the numerous ways the SMART Board was disciplining my thoughts) affects how we maintain and regulate certain possible identities (both our own identities and those of others). This type of work requires a rethinking of the self from the traditional humanist notions of a core and essential self where one progresses through time toward their "real" self, to a subject that is always already implicated in any number of relations of power that constitute possible ways of being in multiple contexts. As I have discussed in the chapter, one such technology of power that is dispersed throughout the social body and works especially in educational discourses—developmentalism— provides a way of limiting these possibilities by dividing selves into categories such as behind, ahead of, or on time, and in the process producing a sense that certain "truths" or knowledge are more proper or normal for us to know. These sorts of practices configure within educational discourses and work on persons in multiple ways such as through the checking of progress *at a glance* with the usage of a SMART Boards or a staged adolescence.

Instead of focusing our efforts on gaining certain knowledges that seem proper to know, thus making us certain types of knowers, Foucault (1985) contends that a focus on power relations and the field of possible actions, responses, and resistances that necessarily open up as a part of these everyday relations is one way in which the important work of "get(ting) free of oneself," or thinking differently than one is currently thinking, can be done (p. 8). This type of freedom is not an essential kind, where one might be free from all requirements or standards that might require one to know how to implement technology appropriately, for instance, or utilize developmentally responsive practices. Instead these freedoms exist as a part of multiple struggles (Foucault, 1982) during the moments when we, for instance, catch ourselves wondering about our own status as a moron in relation to our knowledge about a technological device. Or find ourselves reminding our students that just because they are of a certain age and within a certain stage, they should be able to maintain a certain level of maturity in all situations. It is in these moments where new possibilities of action occur and we might begin the slow process of reimagining other ways of being in educational settings.

What if, for instance, in those moments where we catch ourselves worrying whether a student is "behind" because they have not yet developed as we might expect, we instead allow ourselves to wonder how each young adolescent is already contributing to the learning community in ways we might not have noticed? What if the classroom teacher who is using a chalkboard instead of a SMART Board is seen not as developmentally delayed, but as making an instructional choice? What if, instead of assuming those concepts that seem most simple and enjoyable are also the most effective in meeting our needs, we began questioning why we think they are so simple, what we enjoy about them, and how we assume they make our lives more effective? What new possibilities for *being* might be possible if we take up Brown and Saltman's call (2005) to open dialogues about that which seems the most natural?

References

Brown, E. R., & Saltman, K. J. (Eds). (2005). *The critical middle school reader.* New York: Routledge.

Dreyfus, H. L., & Rabinow, P. (1982). *Michel Foucault: Beyond structuralism and hermeneutics* (2nd ed.). Chicago, IL: The University of Chicago Press.

Ergo in Demand, Inc. (2000–2009). *About blackboards—Blackboard technology and chalkboard history advances.* Retrieved March 28, 2009, from http://www.ergoindemand.com/about_chalkboards.htm.

Foucault, M. (1978). *The history of sexuality: Volume 1. An introduction* (Robert Hurley, trans.). New York: Vintage Books. (Original work published 1976.)

———. (1979). *Discipline and punish: The birth of prison* (A. Sheridan, trans.). New York: Vintage Books. (Original work published 1975.)

———. (1982). The subject and power. In H. Dreyfus & P. Rabinow (eds), *Michel Foucault: Beyond structuralism and hermeneutics* (pp. 208–226). Brighton, Sussex: Harvester.

———. (1985). *The history of sexuality, Vol. II: The use of pleasure* (R. Hurley, trans.). Harmondsworth: Penguin.

Knowlton, N. (2006a). *Interactive whiteboard usability.* Retrieved July 30, 2008, from http://www2.smarttech.com/NR/rdonlyres/0547408B-8B62-4C32-A649-53E5DE753507/0/2006InteractiveWhiteboardUsabilityUSNPUpdated07.pdf.

———. (2006b). *Guide on the side.* Retrieved July 30, 2008, from http://www2.smarttech.com/NR/rdonlyres/D11377F8-D2A9-48E3-A91B-17F9666D1F6E/0/2006GuideontheSideUSNPUpdated07.pdf.

———. (2006c). *21st-century students and skills.* Retrieved July 30, 2008, from http://www2.smarttech.com/NR/rdonlyres/260FD482-59B1-4-BB8-AB6A-D0B93DB0D4F1/0/200621stCenturyStudentsandSkillsUSNPUpdated07.pdf.

———. (2007). *The 21st century classroom.* Retrieved July 30, 2008, from http://www2.smarttech.com/NR/rdonlyres/0DB04732-B65B-4F72-BBF1-84AA0B8B3AD1/0/200721stCenturyClassroomFinal.pdf.

———. (2008). *Realizing a teaching and learning dream.* Retrieved July 30, 2008, from http://www2.smarttech.com/NR/rdonlyres/2E7E939E-4E58-4A98-A714-F8F3B98F2902/0/RealizingATeachingDream_Dec3ML.pdf.

———. (2009). *Engagement and assessment—a student's perspective.* Retrieved July 30, 2008, from http://www2.smarttech.com/NR/rdonlyres/AF8A4732-3B9E-45DC-9E58-82F05749FC0C/0/2009_EngagementandAssessment.pdf.

Lesko, N. (2001). *Act your age: A cultural construction of adolescence.* New York: Routledge.

Rabinow, P., & Rose, N. (2003). Introduction. In P. Rabinow & N. Rose (eds), *The essential Foucault: Selections from essential works of Foucault, 1954–1984* (pp. vii–xxxv). New York: New Press.

Saltman, K. J. (2005). The social construction of adolescence. In Brown & Saltman (eds), *The critical middle school reader* (pp. 15–20).

Slate. (2009). Retrieved March 28, 2009, from http://en.wikipedia.org/wiki/Slate_(writing).

CHAPTER 10

A Critical Perspective on Human Development: Implications for Adolescence, Classroom Practice, and Middle School Policy

Enora R. Brown

INTRODUCTION

This chapter examines ways in which *critical* and *traditional* understandings of human development and the concept of adolescence inform school policy, and influence teachers' practices and pedagogical "responsiveness" to the learning "readiness" of middle school youth. It begins with a critical perspective on human development, embracing the multiple embedded sociocultural, historical, biological, relational, and psychological processes that constitute human growth and change throughout the lifespan and across generational time. This *critical view* counters the *traditional view* that human development is primarily a naturally unfolding biological process, tangentially influenced by culture, "proceed[ing] toward a unique desirable endpoint of maturity...[along] a single developmental trajectory...[as] a linear cultural evolution...from primitive to 'us'" (Rogoff, 2003, p. 18). The critical perspective is based on the assumption that development is mutually constituted by dynamically interwoven *cultural and biological processes*, occurring through interpersonal and societal relationships, communities' historical practices and artifacts, power relations and institutional hierarchies, and meanings ascribed to these processes. Assumptions within traditional and critical perspectives have implications for views on the "nature" of learners, "role" of teachers, and overarching educational goals.

This chapter considers traditional and critical understandings of the adolescent construct, concepts of "student readiness" and "teacher responsiveness," and their implications for educational policy and educative relations with middle school youth. It discusses some unexamined assumptions in traditional conceptions—highlighting the need to rethink notions of "developmentalism," utilize critical lenses, enabling educators to question normalized developmental views, and to envision complex youth constructs and new pedagogical practices in schools. Four chapters—that push against the normative grain of developmentalism—in this book, by Bridges-Rhoads, Harrison, Hughes, and Vagle, are discussed. They examine student readiness and teacher responsiveness in critical literacy, essentialized racial constructs of adolescent identity, linear development as a technology of power, and assumptions about sexuality in adolescent female identity constructs—all revealing innovative ways of thinking and being in middle schools. This chapter highlights the power and necessity of envisioning new developmental and educative theories and practices toward democratic aims. It argues for critical understandings of human development to transform educators' views of youth, their ability to promote a wider array of life options, and inhibit proscriptive life trajectories for middle school youth in divergent communities.

A Critical Understanding of Human Development

What is human development? Traditionally, human development has been conceptualized as an internal biological process—as the natural unfolding of predictable stages of growth and change, and the emergence of the individual child's social, emotional, intellectual, and physical characteristics—peripherally shaped by social relationships and essentialized "cultural differences." The paradigm of human development positions culture as an external influence, rooted in the rise of nineteenth-century industrialism and colonialism, as European Americans "made sense" of observed differences between themselves, the human norm, and those conquered and colonized peoples within and outside of the U.S. borders—the culturally aberrant. The discipline of human development reflected the ascendance of positivism, Cartesian logic, and efforts to model the "soft sciences" (i.e., studies of humanity) after the "hard sciences" (i.e., studies of the inanimate world). Though few debate the biological dimension of the development of humans or the predictability of distinct features of the human species (e.g., walk upright, talk, think), the traditional view tends to foreground biology and genetic heritability, marginalizing the centrality of culture and constellation of personal and societal relationships that are integral to, not supplemental to, individual and cross-generational processes of change and growth.

Three assumptions distinguish the traditional from the critical perspective. First, human biology and culture are viewed as independent internal and external entities. Second, universal developmental goals that exist across communities are considered biological, while variations are considered cultural. Third, culture is viewed as a force exerting influence from the outside "onto the otherwise generic child" (Rogoff, 2003, p. 41). These assumptions obscure the range of *biological differences* within societal communities and *cultural similarities* across the human species (p. 64). They undergird the Eurocentric belief in one "civilized" developmental trajectory and "universal" set of developmental milestones for humanity, promoting narrow codified notions of "normalcy," "cultural difference," "abnormalcy," or "deviancy." Comprehensive interdisciplinary understandings of the cultural nature of human development challenge the traditional view and interrogate the role of meaning-making in developmental processes.

What is a critical perspective on human development? There are three salient issues. First, human development, by definition, is a cultural process, occurring through reciprocal interpersonal and societal relationships (Elliott, 2001; Rogoff, 2003). Second, a comprehensive, complex view of culture includes societal institutions and the circulation of shared meanings and power that normalize developmental processes and knowledge produced (Hall, 1997; Ogbu, 1988). Third, local and global structures of social inequality and historical power relations are embedded in developmental processes and discourses (Cannella, 1997; Holland & Lave, 2001; Spencer & Tinsley, 2008). Through different disciplinary theories, these issues address the synergy between biology and culture, psychological and social dynamics, historical and current institutional structures, individual and societal processes, meaning-making and cultural practices, and materiality and ideology that constitute human growth and change processes. A critical view embraces the individual participants' agency in cultural communities and our inherited capacities as humans. It considers relations of power that contribute to disparities in development and justificatory dominant discourses. Educators' attention to these issues may facilitate individual and social transformations to promote maximal possibilities for all youth, as the following sections show.

THE ROLE OF CULTURE IN DEVELOPMENT

Rogoff's work (2003) is grounded in sociocultural historical theory and provides considerable insight into the role of culture in development. She states, "Humans [are] biologically cultural" (p. 62).

> Human development is a cultural process. As a biological species, humans are defined in terms of our cultural participation. We are prepared by both our cultural and biological heritage to use language and other cultural tools to learn from each other... People develop as participants in cultural communities. Their development can be understood only in light of the cultural practices and circumstances of their communities—which also change... humans [are] biologically cultural. (Pp. 3–4)

Rogoff asserts that human beings are inherently biological *and* cultural, that biology and culture are inextricably bound together and mutually constitutive in developmental processes. With intentionality, she explicates the significance of culture and foregrounds its pivotal role in development. Through *guided participation* (ZPD) in sociocultural activities in cultural communities and our requisite engagement in interpersonal and societal relationships, individual (ontogenetic) development builds on our cultural heritage as a species (phylogenetic) development. Simultaneously, as individuals create new cultural practices and tools that generate change in cultural communities, they contribute to phylogenetic development. Thus, cultural practices, tools, and developmental goals, integral to individual biological change and human societies' preservation, are transformed across generations as they "contribute to biological evolution" (p.66). Rogoff's view removes artificial boundaries between biology and culture, the individual and society, and generational cultural inventions and human evolution, and highlights their permeability in constructing and transforming each other in developmental processes. Differential goals of *independence and separation* versus *interdependent autonomy* are examples of the biologically cultural nature of humans and cultural nature of human development.

In the United States, *independence and separation* are revered qualities, developmental goals, and indicators of maturity that are embodied in cultural practices and tools for infants, school-age children, and adolescents. For infants, there are separate infant-parent carrying patterns, solitary sleeping arrangements, infant seats and cribs (Rogoff, 2003; Shweder, 2003). For school-age children, there is a press for exclusive assessments (e.g., tests, of individualized, competitive versus collective peer group). For adolescents, there are expectations for "sturm and drang-plagued" (i.e., storm and stress-plagued) youth to distance themselves from parents, eventually "leaving the nest" after high school or college. The individual-focused developmental aim toward independence cannot be separated from the United States' historical formation and philosophical traditions (i.e., the colonies' struggle for independence from England, settlers' "trailblazing" adaptation to climactic and geographic conditions, and "pioneering" territorial conquest of Native American land through westward expansion). Industrial production, individual-based meritocracy, private property, and entrepreneurship were foundational in the nation's ascendance, and were rooted in society's cultural devotion to individualism, privacy, competition, and independence, which permeate cultural goals of human development in the United States (Brown, 2008; Spring, 2007).

Interdependent autonomy (i.e., group coordination and personal freedom) is a different developmental goal in other cultural communities, which is commensurate with their historical efforts to sustain life (Rogoff, 2003). In some African and Polynesian communities, infants are bound to their caregivers' bodies with clothing wraps and are carried by them twenty-four hours a day through work and leisure. In impoverished communities worldwide, children do not sleep alone because separate sleeping arrangements are neither preferable nor possible. In some agrarian societies (e.g., in the Global South) lacking basic necessities or "surplus" resources, collective work is necessary and desired to produce goods for the survival of all community members. In some European countries (e.g., Italy), across social class, young adults stay with their family of origin until they create their own families. The cultural practice of integrating or segregating youth, or promoting independence versus interdependence within communities, has significance. There are implications for infants' self-transformation within interdependent relationships; for school-age children's individualism and solitary work versus collectivity and cooperative learning; for cognitive evaluative assessments and tracking; and for adolescents' eventual pursuit of independent, self-sufficient living—as parents anticipate the "empty nest." Other cultural communities expect young people to live at home and to become successful, contributing societal members by building reciprocal familial relations (Rogoff, 2003).

Divergent goals and cultural practices have biological implications for "wired" human qualities and constructions of normalcy. Different infant-caregiver sleeping patterns may be related to infants' biological rhythms (e.g., breathing patterns, neurological activity, Sudden Infant Death Syndrome). Though all humans are biologically prepared to walk upright, the onset time for African and European American children differs. African children—who are carried by their caregivers daily in communities without stairs—absorb ambulatory kinesthetic rhythms, walk a few months earlier, and climb stairs later than European Americans (Rogoff, 2003). Textbooks define African children's earlier ambulatory ability as "precocious" (i.e., deviating from American children's "normalized" onset time, no longer "behind" Africa). The West's press for "social efficiency" and numerical measurement characterizing early industrialism's scientific inquiry prompted age-norms, grade-based milestones, time-driven classification systems, and cultural definitions of "on time," "advanced," and "lagging behind," all of which have psychological and social implications for the self-other views that youth share, internalize, and negotiate. Youth make experiential sense of their relative age and grade-based social position within and across communities and their ascribed mental health and social maturity level based on the timely attainment of developmental milestones toward adulthood (Rolon-Dow, 2004; Spencer et al., 2001).

Common goals of development, attained and expressed through different cultural practices, are exemplified in the *reciprocal emotional attachment* children develop with their caregiver(s). Efe children in the Democratic Republic of the Congo are nurtured and comforted by multiple caregivers, develop close attachments to their mothers, and express less stranger and separation anxiety than children of white Bostonian mothers, who rely on single caregiver arrangements (Tronick et al., 1987). Though *social experience* enables children's inherited maturational capacity to talk, variations exist in the value of narrative versus descriptive speech among some African American and European American children. Cultural communities in South Asia, Iran, Africa, Palestine, South America, the West, Afghanistan, and indigenous Aboriginal Australia envision and pursue different developmental goals and paths to common goals. These goals are culturally and biologically directed toward specific *and* general conditions of social life, imbued with cultural meanings to support the human quest to survive and thrive.

These examples illustrate the permeability of biology and culture, the individual and society. Anthony Elliott (2001), a British social theorist, states: "Social processes in part constitute, and so in a sense are internal to the self. Neither internal nor external frames of reference should be privileged" (p. 5). He synthesizes psychological and social dimensions of life, providing a conceptual lens to envision how societal values and institutional dynamics reciprocally pervade our developmental concepts and goals, community and familial relationships and practices, and individual behaviors and sense-making processes. Inquiries into the cultural nature of development reveal current limits of and future possibilities for knowledge produced through rigorous, interdisciplinary theory and reflexive, *emic and derived etic* research approaches (Rogoff, 2003).

CONCEPT OF CULTURE: WAY OF LIFE, SHARED MEANINGS, AND POWER

The second issue that contributes to a critical human development perspective is a comprehensive, complex view of culture, including societal institutions and the circulation of shared meanings and power. John Ogbu's (1988) and Stuart Hall's (2005) work provide insights. *Culture* is often relegated to customs (e.g., food, rituals, art, music, language across varied ethnic and racial communities), with little attention to meanings embedded in practices, their historical significance, or contribution to human survival in particular ecological contexts. The dominant culture arbitrarily defines these differences as refined versus backward or "high" versus "low" or popular culture, for example, "the classics" versus multicultural literature, classical music versus jazz or hip hop, art by "the masters" versus "folk art." *Culture* is used as a categorical property of individuals, a static conglomeration of immutable qualities, that "inherently" defines a group, for example, nationality, ethnicity, race, and determines their progression along "the" developmental trajectory (Rogoff, 2003). Ogbu's comprehensive anthropological perspective (1988) addresses not only people's ritual behaviors, customs, and artifacts, but also *dimensions* defined as the *imperatives of culture*, embodied in social, economic, political, and religious structures and institutions of society. He states:

> Culture is a way of life shared by members of a population. It is the social, technoeconomic, and psychological adaptation worked out in the course of a people's history. Culture includes customs or institutionalized public behaviors, as well as, thoughts and emotions that accompany and support those public behaviors. It includes artifacts— things people make or have made that have symbolic meaning.

> Particularly important is that the definition of culture includes people's economic, political, religious, and social institutions—the imperatives. These imperatives form a recognizable pattern requiring competencies that guide the behaviors of members of the culture fairly predictably...Culture...ensur[es] the physical survival of children [through] the acqui[sition of] the cultural attributes of their society. (Pp. 11–12)

Culture, the way of life of a people, includes their relationships, institutions, behaviors, traditions, and worldviews, their developmental goals and processes, and revered forms of maturity that embody inherited historical practices, meanings, and artifact use, forged over time for human survival. The imperatives of culture—social and religious institutions, political and economic structures—pervade multiple levels of human interaction and life, and are often omitted from developmental considerations in the West. Social, institutional, and societal processes at the macro level are viewed as separate and unrelated to individual processes of growth and change at the micro level. In the 1950s, hospitals' standard institutional practice was to take newborn infants from their mothers at birth and place them under nurses' care until feeding time. This cultural practice embodied society's valued cultural investment in separation and independence as developmental goals, had biological and socio-emotional consequences (e.g., for reciprocal mother-infant bonding), and conveyed meaning about the relative importance of mothers' emotional and biological care for her child versus custodial care by medical personnel. As imperatives of culture, institutional practices are integral to human development. Ogbu's anthropological view (1988) delineates various cultural dimensions. The cultural studies perspective focuses on culture as a meaning-making, power-imbued process. Though often placed at odds, both contribute to understanding the complexity of culture as a consequent and determinant of human life.

Stuart Hall (1997) asserts that meaning-making saturates society. He theorizes a *circuit of culture* through which *shared meanings* about the animate and inanimate world are produced, circulated, consumed, and regulated within power relations. He states:

> Culture...is not so much a set of *things*...as a process, a set of *practices*. Primarily, culture is concerned with the production and the exchange of meanings—"the giving and taking of meaning"—between the members of a society or group... "making sense" of the world, in broadly similar ways...cultural meanings are not only "in the head." They organize and regulate social practices, influence our conduct and

consequently have real, practical effects. (P. 2; emphasis in the original)

Hall explains that shared meanings, created and negotiated through language (i.e., primary representational system for people's concepts, ideas, and feelings), through discourse, social practices, and interactive dynamics among people and institutions have real effects. The five foci in the circuit of culture—representation, regulation, identity, production, and consumption—illustrate this. Through representation's multiple forms (e.g., media, photography, artifacts, technology), meanings are produced, consumed, and circulated throughout culture. Though the meanings are not "fixed," unitary, or inherent signs of people, objects, or events, they are instrumental in producing and assigning identities around dimensions of difference, and are regulated and structured in particular patterns. They are regularized and naturalized through social interactions, as institutional, governmental, and societal policies and practices. These "real" meanings, social identities, and practices are "reproduced," consumed, and (re)presented as if "true." Power relationships determine who assigns differential meaning (e.g., to people, events). As symbolic power circulates throughout meaning-making processes, the knowledge produced marks, classifies, "naturalizes," and "help[s] to set the rules, norms, and conventions by which social life is ordered and governed" (Hall, 2005, p. 297). Not confined to the local, the circuit of culture operates through economic and political processes of globalization, wars of domination and "democratization," and exchanges of science and the arts that transport, hybridize, and transform cultural practices, social identities, and meanings around the world. Through this discursive process, meanings are attributed to development, producing official knowledge in written, verbal, and visual texts.

Thus, *culture is productive*. For example, stereotypes codify meanings from exaggerated reductions of persons to simplistic traits, which are normalized, contested, influence conduct, policies, and institutional practices (e.g., the struggle against systemic racism and gender inequality). Hall (1997) states: "[C]ulture depends on giving things meaning by assigning them to different positions within a classificatory system" (p. 236), and has psychological and social, ideological and material effects on individuals and society. A critical perspective requires inquiry into cultural practices' assigned and contested meaning (e.g., valuation of the developmental goals of separation and independence), and into various discourses of development (e.g., the inclusion or exclusion of culture and power in the discipline of human development).

Human Development: Structures of Social Inequality and Historical Relations of Power

The third issue contributing to a critical view of development is the pivotal role of *local and global structures of social inequality and histori-cal relations of power*, embedded in the developmental processes and discourses and operating through the imperatives of culture—social, religious, political, and economic institutions. Structural power rela-tions pervade histories of social inequality, oppression, rebellion, and privilege that are inherited and embodied in current stratified social arrangements and organized around race, class, ethnicity, language, gender, and sexuality. They include dominant and nondominant dis-cursive values and practices that reciprocally support or challenge the culture's institutional hierarchies and structural relationships.

Though Elder (1974) addresses history's *influence* in development, Holland and Lave (2001) theorize its significance at the point of local practice. In their concept, *history-in-person,* histories of social inequal-ity operate fully in the present, via daily interactions and cultural practices. They state:

> The energy of enduring struggles—carried out for and against societal institutions and discourses that disproportionally distribute symbolic and material resources to favored racial, ethnic, class and gendered groups—has been realized in local practice and brought from there into the intimate...History is made in person, registered in intimate identities [and] in institutions. (Pp. 13 & 17)

The authors assert that histories of enduring struggles against inequality are embodied in people—enacted and negotiated locally through participants' interactions. For example, African American and Latino youth's overrepresentation in special education is legiti-mized by technologies of power (i.e., tests, tracking systems) and seems "natural," given historic beliefs about their inferiority and "culture of poverty," and concomitant beliefs in white youth's intel-lectual and moral superiority—evidenced by their "natural" place-ment in advanced placement (AP) classes. Parents, students, teachers, and others contest institutional practices and discourses in K-12, higher education, federal and state institutions that systematically identify, classify, and label certain students as "behavior disordered" and "intellectually challenged." Their opposition to English-Only policies, sanctions against youth's limited English proficiency, and failure to learn English quickly and assimilate are equated with "cognitive deficiencies," low achievement, special needs, and unpatriotic refusal to abandon their first language. They oppose the profound disregard for the socio-emotional, psychological, intellectual, and familial and relational significance of humans' communicating historically shared meanings in their first language, and the implied superiority of English and the false promise that it, in its role as a "common language," produces equality and ends oppression (Macedo, 2005). These local struggles in schools, courts, and government reenact inherited struggles for equality against racism and colonialism. Similarly, the "girl problem" in math implicates girls' limited aptitude and interest in mathematical reasoning. This genetic-oriented view ignores pre-twentieth-century history of gender inequality, when African American and Native American women, neither free nor citizens, were marginal to society, and free white women were excluded from certain educational, political, and economic pursuits, manifest in current challenges and inequalities for girls in school and mathematical career pursuits.

Histories of inequality and power relations are rarely addressed in explanatory paradigms of development, except for a few exemplars (Burton et al., 1996; Cannella, 2004; Spencer & Tinsley, 2008). Under the guise of "objective" inquiry, they are silenced as "political" issues unrelated to development. This silence naturalizes reductionist biological explanations for developmental patterns by race, ethnicity, class, gender, sexuality, and obscures the sociohistorical, political, and economic sources of difference (Bowman, 1990). Thus, cross-generational poverty and unaffordable nutritional food are rarely cited as sources of malnutrition, developmental delay, or unhealthy dietary patterns (e.g., high fat, carbohydrate foods); biological explanations for racial and ethnic achievement disparities range from brain size and Social Darwinism's "survival of the fittest" to ascriptions of "cultural deficiency" and arrested development (Gould, 1996); the declining pubertal onset age is blamed on youth's sexual penchant rather than U.S. culture's fat intake.

Omission regulates discourses and "legitimate" knowledge produced about development's classificatory categories, related school policies and practices, social identities constructed around difference, and socio-educational experiences of youth as they negotiate self-meanings in relation to others. The cumulative history of social, economic, and political inequalities passed down generationally and reproduced in current social arrangements is ignored, and marginalized youth's adaptations to economic and racial or ethnic oppression are rendered "pathological" or poor decisions. Though all youth and adults need and seek organized opportunities to socialize with peers, poor youth of color who play neighborhood basketball and aspire toward sports careers are often disparaged for "choosing" basketball over the debate team "choice" by middle class and white youth. Castigated for low aspirations and naivete in seeking success through sports versus academics, their pursuit of requisite peer interactions—in abysmally poor communities—and adaptive "aspirations" that reflect an "internalization of objective probabilities" goes unacknowledged (MacLeod, 1995). In 1893, paleobotanist Lester Frank Ward opposed Social Darwinism stating: "The denizens of the slums are not inferior in talent to the graduates of Harvard College...forced...to mak[e] the best use they can of their native abilities" (as reported in Kliebard, 1995, p. 23). Social forces—savagely unequal schools, debilitating poverty, and systemic racism—are *delinked* from the forced choices afforded poor youth. Inherent individual or "cultural" deficiency discourses explain different life trajectories, rather than economic and educational disparities that structure unequal class, racial, and gender-defined career and life options. They distort views of "who-is-normal-and-why," undercut understanding developmental similarities and differences, hegemonically enforce homogeneity, marginalize non-dominant cultural practices, constrict the full range of cultural practices and developmental adaptations for human survival, and inhibit possibilities for changing cultural practices that promote and maintain deleterious effects on youth's learning, development, and well-being.

Similarly, the discursive stratification of global developmental differences are delinked from histories of colonization and imperialism. This positions the "Third World" and its diasporic, colonized members at the bottom of the developmental continuum and positions the "First World," its global leaders, and colonizing imperial powers at the top. The "underdeveloped" represent the stymied growth of that sector of humanity at the "evolutionary dead end." The "developed" represent the epitome of progress, who made the "evolutionary leap forward" for the species, and are best equipped to "develop" others (Schech & Haggis, 2000, p. 18). This official discourse promotes a racialized, class-based developmental progression from primitive to civilized, backward to modern, savage to refined, and deficient to normal. It is woven into the economic and political fabric of national and global policies, as meanings define who people are, where they belong in the social order, and what they deserve.

Developmental knowledge reproduces social inequality, or contests veiled developmental constructs—borne of social inequality.

The issues in this section contribute to a critical understanding of human development. Interdisciplinary theories by Rogoff, Elliott, Ogbu, Hall, and Holland and Lave addressed the role of culture in development, a comprehensive, complex concept of culture (i.e., institutions, power-imbued meaning-making processes), and structural inequalities and relations of power as integral to developmental processes and discourses. Thus, human development is a culturally imbued process that occurs across interdependent domains—physical, social, emotional, cognitive, psychic, and spiritual—through dynamic interpersonal (family, peer, community) and social relationships (institutional, societal). It includes the reciprocally intertwined processes of individual growth and change throughout the human lifespan and the evolutionary growth and change of the human species across historical time, shaped by power-imbued relationships and meaning-making processes. The next section examines traditional and critical assumptions about adolescence, developmentalism, student readiness, teacher responsiveness, and their implications for middle school practice. In conclusion, four volume chapters that engage new thinking about adolescents and new principles of development for teaching-learning processes are discussed.

IMPLICATIONS FOR DISCOURSE ON ADOLESCENCE, MIDDLE SCHOOL POLICY, AND CLASSROOM PRACTICE: TRADITIONAL AND CRITICAL VIEWS OF ADOLESCENCE

Traditional biologically centered views of adolescence are grounded in G. Stanley Hall's *Adolescence* (1904). His Darwinist assertion was that ontogeny recapitulates phylogeny, as white adolescent males relive and expend their ancestral primitive savagery as inoculation against effeminancy—to prepare them to be "super-men," who create a "more perfect" advanced civilization, assuming they do not succumb to perversion, hoodlumism, academic-induced neurasthenia, or the pubertal temptation of masturbation. White boys could build on their parents' new evolutionary qualities, African Americans remained child-like, frozen in adolescence based on their ancestors' limited intelligence as a "lesser race," and girls, biologically incapable of inheriting genius, could jeopardize their inherent maternal potential with overeducation. Because the age of religion and white boys' sexual maturity converged with the rechanneling of their sexual energy into education's "repetitive drill and discipline," and the "purifying" variability of their racial recapitulation, white boys embodied civilization's future hope (Bederman, 1996). This view undergirds current traditional perspectives. Though adolescence was equated with puberty, a critical view reveals that it was a sociocultural product of confluent processes (Brown & Saltman, 2005).

The late nineteenth- and early twentieth-century construction of adolescence was grounded in the *imperatives of U.S. culture* (i.e., social, economic, political processes of capitalist industrialization and colonization, proliferate academic inquiry, scientific discovery and longer life, U.S. consolidation of power relations, Protestantism, Victorianism, and Progressive Era universal public education), through which requisite meaning-making processes categorized pubescent youth as a distinct age-group. Scientific advancement to electro-mechanics abruptly transformed an agrarian slave-based economy to an urban industrial one, precipitated social upheaval throwing the destitute masses off the land, created huge demand for northern assembly-line wage-laborers, and generated a massive influx of European immigrants. Searching for work, immigrants, rural poor, and dislocated youth populated burgeoning urban centers. Youth provided cheap labor, but industrial work was the purview of adults, and hordes of idle youth roved the streets.

The elite had major concerns about urban blight, social disintegration, timing of puberty and marriage, sexual depravity, "Americanizing" immigrants, cultivating hard working loyal citizens, promoting Protestant ethic-based morality—in sum, building a unified nation and strong economy. Escalating middle class Victorian preoccupation with civil respectability, repressed sexuality, and self-denial was directed toward harnessing massive societal changes and controlling the masses and reflected G. S. Hall's sentiments. Rising ethnocentrism and the "threat" of unbridled sexuality, "savagery" of youth and the "lesser races" (i.e., "ethnic" immigrants), "colonials," Native Americans, African Americans, Mexican Americans propelled *sexuality and race* to the center of Hall's adolescent construct (Moran, 2000).

Compulsory education (1852–1918) became the means to ideologically consolidate the nation around common values, (e.g., promptness, individual responsibility, self-restraint, loyal participation in government, and adherence to Protestant morality and work ethic), consonant with the needs of profitable industrial capitalist production. Hall's developmental rationale provided the template for junior high school (1910), the first middle level education—to harness adolescent energies, warehouse "wandering bands" of youth, and redirect and indulge white boys' proclivities for their future role in civil society. Educational "social efficiency" mimicked Industrial Taylorism's precise assembly-line piecework (1911), with specialized, standardized differentiated curricula (i.e., classical and vocational) to "tailor" students to their gendered, race, class station in industrial life, and inculcate them with American values (Brown, 2008).

Sweeping societal changes and accompanying "moral" and social anxieties created conditions for constructing adolescence. Capitalist industrialization reconfigured the economy, dislocated the rural populace, generated radical migration and immigration patterns, launched the formation of an industrial working class, and consolidated economic and political relations of power in the new republic. As medicine and scientific discoveries prolonged life, the earlier onset of puberty and delayed marriage created "space" in the lifespan for youth to "prepare" for adulthood. Sharp age distinctions between adolescents and adults were solidified by laws on school attendance, voting, child labor restrictions, contract signing, property ownership, driving, and new organizations emerged to build male character, socialize, control, and segregate youth [e.g., Boy Scouts (1910), Settlement Houses, YMCA/YWCA, Juvenile Justice System (1910), playgrounds, and institutionalizing a distinct youth culture (Lesko, 2001)]. Hall founded the American Psychological Association (APA) in 1892, which institutionalized child study—producing academic knowledge backed by "science" that "made sense" of these societal changes, and named and assigned groups to their position on the social hierarchy. Official texts discursively created, naturalized, and *structured social identity by age* by projecting society's sexual anxieties and forcing displacement of youth onto adolescents as inherent age-and puberty-driven characteristics (e.g., vagrant, impulsive, delinquent, tempted by peers and sex).

Despite sociocultural processes, biological explanations prevailed in developmental discourse. The discipline of psychology and public education's social efficiency practices normalized attributions of societal dynamics to individual biology, naturalized the discourse, and institutionalized adolescents' social role in society. Operating through *the circuit of culture*, societal values and educational, economic, political, ideological processes and power relations produced, consumed, structured, and regulated meanings about youth—forging hierarchical social identities of adolescent and adult that support a stratified society based on age, race, gender, and class.

CONTINUITY OF THOUGHT AND POLICY INTO THE TWENTY-FIRST CENTURY

Though Hall's adolescent construct and rationale for "drill and discipline" middle level education are over a century old, continuities in underlying assumptions about youth exist in official texts and policy documents. Despite twenty-first-century changes in language and commentary on his absurd theories, discursive kernels of Hall's thought prevail: (i) the centrality of puberty and sexuality, (ii) normalization of white middle class patterns and pathologization of youth of color, and (iii) disparaging terminology about youth, projecting societal issues onto adolescent biology. Though some have questioned the biological and racial meanings or elucidated relations of power embedded in the adolescent construct (Bederman, 1996; Lemlech, 2006; Lesko, 2001), puberty and sexuality remain at the hub of the dominant discourse on adolescence and the middle school. William Alexander's conceptualization (1968) of "the emergent middle school" in 1968 was based on a generation of academic child studies, documenting the "nature of adolescence." *Middle school* was designed to meet the developmental needs of adolescents, and replace junior high with a program focusing on "in-between-agers," that is, "pupils between... two stages of maturation, standing at the threshold of puberty..." Primary concerns about this transition period were bodily awkwardness, "onset of menstruation... nocturnal emissions," "new erotic sensations," "social skills...with...the opposite sex," "changes in the activities of the peer group." Secondary were changes in parental relations, self concept, views of right and wrong, and thinking modes (pp. 159, 164). All were central issues in the National Middle School Movement's reformist aims.

Recent texts tentatively disavow, without relinquishing, pivotal ideas from the past, which disparage youth and racialize development. Since 2001, some textbooks modestly distanced themselves from Hall, embracing the idea of "modified storm and stress" for some youth in the United States. Some texts imply that youth are different in "traditional pre-industrial cultures," which maintained a "stable way of life," and do not "value change" (i.e., are transfixed in time, underdeveloped); others take a pluralist approach, focusing explicitly on culture and difference (Arnett, 2007; Rice & Dolgin, 2005). Pervasive discourses still normalize middle class whites in the West, and pathologize youth of color and the Global South. Most texts focus on "adolescent problems," refer to youth as transescents, "like scraggly cygnets", who are "hypocritical" and "preoccupied with self," have "turbulent emotions," unusual, drastic, erratic, and inconsistent behavior (Arnett, 2007; Simpson, 1999; Wiles & Bondi, 2001). While adhering to the traditional view, some texts highlight stereotypes about adolescents, but reproduce racial ones, focus on individual, cultural, and gender differences, and tend to social-emotional and other youth needs to prevent imminent perils (Berk, 2004; George et al., 1998). While cogent critical work has revealed factual distortions in mainstream media and policy scapegoating youth violence, sexuality, and drugs and verifying racial, gendered, and class representations (Dyson, 2004; Giroux, 2009; Males, 2005), current progressive efforts to "humanize" youth and their "developmental needs" unwittingly maintain the conceptual kernel of adolescence, subtly relegating youth to an undeveloped, almost fully human state of adulthood.

Traditional discourse on adolescence underlies major middle school policy reform. Carnegie Council's *Turning Points* (Carnegie Council on Adolescent Development, 1989), the corporate sector's clarion call, launched the Middle School State Policy Initiative (MSSPI), presaging NCLB's privatization and standardization of public education. Grounded in Hall's early admonitions about adolescent temptations that threaten society's progress, it warns that the adolescent "period of trial and error" jeopardizes their preparedness for future work in a global economy and for responsible, ethical citizenship—due to sexual promiscuity, pregnancy, STDs, unhealthy babies, drugs, poor decision-making with confused peers, isolation, disengagement, and school dropout. The subsequent *Turning Points 2000* (Jackson & Davis, 2000) recalibrated middle school education to be in sync with NCLB's 2001 passage. Seven recommended principles focus on standardization, instructional methods, student performance, and the acquisition of discrete information and skills, based on "what students should know and be able to do" (p. 23). The teaching-learning principles mirror Hall's idea (1904/2005) of "inculcation and regimentation" (p. 23), strikingly omitting key elements from *Turning Points* (Carnegie Council on Adolescent Development, 1989), *cooperative learning, critical thinking, and the elimination of tracking*, paving the way for NCLB's middle school reconfiguration. NCLB spurned unprecedented school dropout, pervasive school closings in poor African American and Latino communities, accelerated public education's privatization, and exacerbated the race-class achievement gap and school segregation (Brown, 2003; Lipman, 2003). The Rand Corporation's *Focus on the Wonder Years* (2004), a third policy iteration, challenges the validity of middle school education geared toward adolescents' developmental needs versus their academic standardized test-based achievement (Juvonen et al., 2004).

As in the nineteenth century, schooling is designed to serve corporate interests, institutionalize tracking, direct students or dropouts to their expected station in life, and to, simultaneously, manufacture consent. The message is that school failure, future unemployment, societal degradation, flaws in the national and global economy, and the diminished future of youth pivot on inherent adolescent flaws and parents' and schools' inability to manage them. It projects societal issues onto the youth, blames unemployment on poor education rather than on foundational crises in capitalism, obscures profound social inequalities among youth, and normalizes the "dummying-down," mind-dulling of the population through banking model, test-based education (Freire, 1993; Lipman, 2003). The adolescent discourse a century ago served the societal shift from a rural agrarian to an urban industrial capitalist economy—and so it is today. Maintaining elements of the traditional adolescent construct supports current shifts within global financial capitalism from labor-saving electromechanical to labor-replacing, microelectronic technology, and its adjoining neoliberal agenda by justifying educational reform and requisite social identities.

STUDENT READINESS, TEACHER RESPONSIVENESS, AND CLASSROOM PRACTICE

Traditional discourse on adolescence has implications for student readiness and teacher responsiveness in classroom practice. Hall asserted that adolescents were "ready" in this "golden hour" of plasticity for regimented skill acquisition, and if squandered would result in their profound handicap. Current conceptions of student readiness are based on developmentalism, the view that children mature through an "inherent, 'natural' stage-wise progression." Youth must reach a particular level of social, emotional, intellectual, or physical ability for learning to take place. Attainment of the requisite level allows domain-general learning (i.e., processing across all areas of knowledge as in Piaget's theory), domain-specific learning (i.e., unique processing within each knowledge domain for math, science, etc.), and higher levels of social functioning (e.g., peer relations) to occur. Though the role of social experience in maturation is recognized, most assert that it cannot be "sped up" or substantively accelerated by intervention, despite Piaget's "American question." Teachers should "respond" to students' current readiness to learn a particular

subject matter, and pursue more advanced work once further maturation has occurred (Watson, 1996). Thus, knowing "what to teach and
how to teach it" is based on "developmental assessment," which constitutes "responsive education" (Stevenson, 2002).

Readiness could have applicability, based on close observation and
critical understanding of biologically cultural quantitative and qualitative changes occurring through lifespan relationships. However, it
is fraught with views of youth as passive, unfeeling receptacles, nativ-
ist child views and behaviorist learning theories, "naturalized" institutional expectations and practices, cultural meanings embedded in
power relations, and social inequality. This has generated adverse
retention and compensatory or intervention practices (e.g., school
failure policy, racialized special education placement, abstinence
efforts to control adolescent "nature," and dismantling ameliorative
intervention programs for poor youth). The cultural construct of
readiness—defined as by-and-for-whom—is conditional and raises
questions: What is the role of culture in defining readiness of institutions in regulating its boundaries and of historical power relations in
producing notions of "delayed" or "accelerated" readiness? What are
authentic readiness measures, commensurate teacher responsiveness,
and the "fit" between youth, curriculum, pedagogy, and school culture as informed by traditional or critical discourses on developmental goals and the purpose of education. Four chapters in this volume
relate to these issues in classroom practice.

The chapters by Vagle, Harrison, Hughes, and Bridges-Rhoads
interrogate developmentalism and consider its educational implications for the identities and learning experiences of adolescents. They
offer experiential and theoretically grounded alternatives for our conceptions of and work with youth in schools. Questions, inspired by
each author's work, are posed at the end.

Vagle argues that developmentalism and "developmental responsiveness" are based on the erroneous assumption that "universal" predetermined, age-designated milestones for youth are natural,
unchanging, and immutable. This assumption subjects youth to readiness expectations, based on ascribed, decontextualized processes of
growth and change; it proscribes a limited view of youth for teachers,
consuming them with the rigid task of charting student growth to
"make the grade." Vagle embraces Lesko's (2001) view that growth
and change are "contingent and recursively relational" (i.e., learning
and development are complex processes occurring repeatedly over
multiple contexts and relationships). Because growth and learning
processes are unpredictable, teachers must go well beyond

proscriptive, developmentally appropriate or responsive pedagogy. Using critical literacy pedagogy as an example, Vagle draws on Bakhtin's concept of *dialogism* to illustrate how "appropriate" teaching practices are based on educators' *answerability* to unique, immediate conditions, needs, and interests of students within given discourse communities, *in the moment*. Recognizing the tensions between planning and flexibility and the constraints of standards, he encourages teachers' pedagogy to be relative, unscripted, and unbound by notions of universality, and subject to students' unanticipated dialogue, reflections, and concerns. The inclusion of multiple voices, perspectives, questions and answers constitutes a dynamic, fluid, ongoing teaching-learning process, in which students and teachers discursively share, create, and contest the nonneutrality of meaning and expand the interdisciplinary curricular experience. Vagle's perspective provides insights into rigid constraints of developmentalism's linearity and universality for middle school students and teachers, and explores emergent possibilities when teaching is contextually grounded in relationships, dialogic interchange, and students' unique qualities. Two questions that emerge are: As teachers engage dialogic pedagogical practices with students in their classrooms, how do they assess their practices' "answerability" to dominant versus nondominant societal discourses (Giroux, 2009)? How do pedagogical classroom decisions disrupt or contest teacher and student discourses or practices that reproduce inequality around difference, in order to promote broader democratic aims (Cochran-Smith, 2004)?

Harrison's chapter examines essentialized racial meanings, embedded within the adolescent construct, and their impact on African American youths' racial identity formation. Drawing on W.E.B. Du Bois's concept of double-consciousness, she discusses the dual challenge that African American youth face: negotiating identity amidst internalized representations of the self as African American, as seen through "the Other's" eyes; and negotiating their adolescent "identity" within Hall's racialized, gendered construct. She raises questions about the adolescent construct's social function in society and its implications for African American youth, and incorporates constructivist elements from cultural studies to critique social identities constructed apart from historical relations of inequality. Referencing Gramsci, she suggests that "common sense" hegemonic notions about race support domination and operate in black youth's identity formation processes. As an example, Harrison describes the tensions an African American male faces in negotiating shifting self-meanings from junior high to college. She discusses the internalized

experience of double consciousness and racialized self-other representations, as the student moves from "acting white," in opposing the
underachievement stereotype, to new "ways of being" African
American through his exposure to alternate African American role
models in college history courses, while counter-demonizing self-
constructions by the Other. Harrison's conceptual model of African
American youth's identity construction strives to address the double-
consciousness dilemma they face in negotiating what it means to be
African American and racialized meanings within "white" adolescent
constructs. She asserts the importance of understanding identity formation among youth of color, educators' and society's role in dismantling dominant views and practices, whose racialized meanings limit
options for youth. As Harrison dispels color-blind discourse, she
highlights the subtle but powerful significance of race in developmental discourse, and unique challenges adolescent youth of color
face in negotiating racism. Related questions are: How may the acting
white notion be problematized and challenged, in light of parent
socialization practices, peer relations, and the dynamics of social class
(Spencer & Tinsley, 2008)? How may scholars reconceptualize the
simultaneous, mutually permeating race and age-based processes nav-
igated by youth to support their journey, and how may teachers interrupt social relations that support certain racialized representations of
youth (Carter, 2007)?

Hughes explores challenges adolescent girls face as their bodies
are commodified in the media, while they are being constructed by
adults as asexual beings, "at that age." Her narrative case study of
"dialogue" from a Girls' Forum highlights the discursive process of
identity construction that occurs within and through cultural practices in schools, and the deleterious consequences of denying female
adolescents' sexuality. She addresses the girls' frustrations in encountering Bhabha's "disjunctive present" (i.e., in trying to negotiate
adult expectations for an asexual self) and make sense of themselves
amidst media distortions of the "attractive" female body. Hughes
addresses religious and gendered views from Hall's Victorian era
embedded in current constructions of female adolescent identity.
Referencing Foucault, she notes the productive nature and power of
silencing sex talk, "the forbidden topic," by constricting parent-stu-
dent-teacher communication and learning opportunities. She
addresses power relations operating in dominant culture's constructed gendered, sexual identities. Through teacher-student dialogue, Hughes probes the limits of biological determinism, the role
of peers in framing self-constructions, and the significance of body

image. She suggests that ideological moorings of parents' earlier sexual experiences reemerge during their interactive judgments of girls, which highlights society's tendency to project sex-as-deviant discourse onto youth, while obscuring adult roles in sexualizing children (e.g., child beauty pageants). In redemptive fashion, Hughes illustrates educators' affirming role—providing spaces for girls to talk about challenges of negotiating gender identity, resisting the "moral" asexual discourse, and promoting new gendered meanings for girls through discussions with parents. Hughes contributes to understanding "sexual desire" as a valued dimension of adolescent female identity, distinguishing between intimacy and sex. Two questions emerge: How do social class and race differentially inform discourses and practices within schools/society on female gender identity, and what are the implications for lived experiences of girls of color and for working and middle class backgrounds (Collins, 2008)? How may educators engage gender identity issues through school curricula and practices to foster dialogue, denaturalize dominant meanings, and forge new youth constructions (Fine, 1992)?

Bridges-Rhoads examines the temporal linearity of developmentalism as a technology of power, which imbues cultural artifacts (i.e., The SMART Board) and the adolescent construct with meanings that discipline teachers and students and forge classificatory identities. Using Foucauldian theory, she analyzes her SMART Board (in)experience, relative to others, to address notions of linear progression that promote feelings of being left behind, advanced in pursuit of an achievement marker, and schools' usage of technology to discursively construct and track ability hierarchies and developmental age-appropriateness in schools. She addresses the adolescent concept and age-stage milestones that categorize knowers, constrict youth to "act your age," and shape teacher expectations, practices, and exhortations to youth. Bridges-Rhoads highlights power's productive nature—embodied in artifacts and constructs—as it unobtrusively pervades social groupings, sets norms, constructs identities, and limits ways of being. The SMART Board's imbued power of inherent progress over earlier technologies defines school participants and "ranks" schools. Bridges-Rhoads examines how individuals interiorize their gaze, police themselves, assess their position relative to normalcy, monitor their progress in ascending toward predetermined, time-ordered goals—sans external force—and experience the adverse effects of "tardiness." She challenges traditional views of a core self progressing toward full authenticity, recommending that teachers free themselves and students by denaturalizing binding, regimented procedures of temporal, unidirectional control to see what has been missed and envision new possibilities for youth. Questions that surfaced are: How do artifacts, as technologies of power (e.g., SMART Board), embody material power across class lines (e.g., wealthy and poor schools) (Brown, 2003)? How do educators challenge material inequalities represented by differential technologies in schools, while negotiating concurrent symbolic power of developmentalism's temporal linearity (Lipman, 2003)?

Each author and this chapter's overarching issues contribute to ongoing examinations of human development discourses and processes and generate efforts to change our educative interactions with youth. It is important, throughout our critical inquiries, to consider the implications of our analyses for pervasive structures and discourses within the dominant culture, our classrooms, and other venues. Through our transformative work in education, we should embrace elements of existing human development constructs that "make sense," while simultaneously challenging and discarding meaning-imbued practices that impede democratic aims for students and ourselves as educators across multiple lines of difference.

References

Alexander, W. (1968). *The emergent middle school*. New York: Wadsworth.

Arnett, J. (2007). *Adolescence and emerging adulthood: A cultural approach*. Upper Saddle River, NJ: Prentice Hall.

Bederman, G. (1996). *Manliness and civilization: A cultural history of gender and race in the United States, 1880–1917*. Chicago, IL: University of Chicago Press.

Berk, L. (2004). *Development through the lifespan*. New York: Allyn & Bacon.

Bowman, P. (1990). The adolescent-to-adult transition: Discouragement among jobless Black youth. In V. McLoyd & C. Flanagan (eds), *Economic stress: Effects on family life and child development, New Directions for Child Development, #46* (pp. 87–105). San Francisco, CA: Jossey-Bass Inc.

Brown, E. (2003). Freedom for some, discipline for "others": The structure of inequity in education. In K. Saltman & D. Gabbard (eds), *Education as enforcement: The militarization and corporatization of schools* (pp. 126–151). New York: Routledge.

———. (2008). Education and the law: Toward conquest or social justice. In W. Ayers, T. Quinn, & C. Stovall (eds), *Handbook for social justice in education* (pp. 59–87). New York: Routledge.

Brown, E., & Saltman, K. (2005). *The critical middle school reader*. New York: Routledge.

Burton, L., Obeidallah, D., & Allison, K. (1996). Ethnographic insights on social context and adolescent development among inner-city African American teens. In R. Jessor, A. Colby, & R. Shweder (eds), *Ethnography and human development: Context and meaning in social inquiry* (pp. 395–418). Chicago, IL: The University of Chicago Press.

Cannella, G. (1997). *Deconstructing early childhood education: Social justice and revolution.* New York: Peter Lang.

———. (2004). *Childhood and post-colonization: Power, education, and contemporary practice.* New York: Routledge.

Carnegie Council on Adolescent Development. (1989). *Turning points: Preparing American youth for the 21st century.* New York: Carnegie Corporation.

Carter, P. (2007). *Keepin' it real: School success beyond black and white boundaries.* New York: Oxford University Press.

Cochran-Smith, M. (2004). *Walking the road: Race, diversity, and social justice in teacher education.* New York: Teachers College Press.

Collins, P. (2008). *Black feminist thought: Knowledge, consciousness, and the politics of empowerment.* New York: Routledge.

Dyson, M. (2004). *The Michael Eric Dyson reader.* New York: Basic Civitas Books.

Elder, G. (1974). *Children of the Great Depression: Social change in life experience.* Chicago, IL: University of Chicago Press.

Elliott, A. (2001). *Concepts of the self.* Malden, MA: Polity Press.

Fine, M. (1992). *Disruptive voices: The possibilities of feminist research.* Ann Arbor, MI: The University of Michigan Press.

Freire, P. (1993). *Pedagogy of the oppressed.* New York: Continuum International Publishing Group.

George, P. Lawrence, G., & Bushness, D. (1998). *Handbook for middle school teaching.* New York: Longman.

Giroux, H. (2009). *Youth in a suspect society: Democracy or disposability.* New York: Palgrave Macmillan.

Gould, S. (1996). *The mismeasure of man.* New York: W. W. Norton & Co.

Hall, G. S. (1904/2005). *Adolescence: Its psychology and its relations to physiology, anthropology, sociology, sex, crime, and religion.* New York: D. Appleton & Co.

Hall, S. (Ed.). (1997). *Representation: Cultural representations and signifying practices.* Thousand Oaks, CA: Sage Publications.

Holland, D., & Lave, J. (2001). *History in person: Enduring struggles, contentious practice, and intimate identities.* Santa Fe, NM: School of American Research Press.

Jackson, A., & Davis, G. (2000). *Turning points 2000: Educating adolescents in the 21st century.* New York: Teachers College Press.

Juvonen, J., Le, V., Kaganoff, T., Augustine, C., & Constant, L. (2004). *Focus on the wonder years: Challenges facing the American middle school.* Santa Monica, CA: Rand Corporation.

Kliebard, H. (1995). *The struggle for the American curriculum, 1893–1958*. New York: Routledge.

Lemlech, J. (2006). *Curriculum and instructional methods for the elementary and middle school*. Upper Saddle River, NJ: Prentice Hall.

Lesko, N. (2001). *Act your age!: A Cultural construction of adolescence*. New York: Routledge.

Lipman, P. (2003). *High stakes education: Inequality, globalization, and urban educational reform*. New York: Routledge/Falmer.

Macedo, D. (2005). English only: The tongue-tying of America. In Brown & Saltman (eds), *The critical middle school reader* (pp. 375–384).

MacLeod, J. (1995). *Ain't no makin' it: Aspirations and attainment in a low-income neighborhood*. Boulder, CO: Westview Press, Inc.

Males, M. (2005). "Bashing youth" and "wild in deceit." In Brown & Saltman (eds), *The critical middle school reader* (pp. 121–130).

Moran, J. (2000). *Teaching sex: The shaping of adolescence in the 20th century*. Cambridge, MA: Harvard University Press.

Ogbu, U. (1988). Cultural diversity and human development. In D. T. Slaughter (ed.), *New directions for child development—Black children and poverty: A developmental perspective* (pp. 11–27). San Francisco: Jossey-Bass.

Rice, F. P., & Dolgin, K. (2005). *The adolescent: Development, relationships and culture*. New York: Pearson.

Rogoff, B. (2003). *The cultural nature of human development*. New York: Oxford University Press.

Rolon-Dow, R. (2004). *Urban girls: Resisting stereotypes, creating identities*. New York: New York University Press.

Schech, S., & Haggis, J. (2000). *Culture and development: A critical introduction*. Malden, MA: Blackwell Publishers.

Shweder, R. (2003). *Why do men barbeque?: Recipes for cultural psychology*. Cambridge, MA: Harvard University Press.

Simpson, S. (1999). Early adolescent development. In C. Walley & G. Gerrick (eds), *Affirming middle grades education* (pp. 5–16). Boston, MA: Allyn & Bacon.

Spencer, M., & Tinsley, B. (2008). Identity as coping: Youths' diverse strategies for successful adaptation. *Prevention Researcher, 15*(4), 17–21.

Spencer, M., Noll, E., Stoltzfus, J., & Harpalani, V. (2001). Identity and school adjustment: Revisiting the "acting white" assumption. *Educational Psychologist, 36*(1), 21–30.

Spring, J. (2007). *American education: From the puritans to No Child Left Behind*. New York: McGraw-Hill Humanities.

Stevenson, C. (2002). *Teaching ten to fourteen year olds*. Boston, MA: Allyn & Bacon.

Tronick, E. Z., Morelli, G. A., & Winn, S. (1987). Multiple caretaking of Efe (Pygmy) infants. *American Anthropologist, 89* (1), 96–106.

Watson, R. (1996). Rethinking readiness for learning. In D. Olson &
N. Torrance (eds), *The Handbook of education and human development*
(pp. 148–172). Malden, MA: Blackwell Publishers.

Wiles, J., & Bondi, J. (2001). *The new American middle school: Educating
preadolescents in an era of change.* Columbus, OH: Merrill Prentice Hall.

PART III

Conversations across Fields

Kyunghwa Lee and Mark D. Vagle

In this final section, two cross-field conversational chapters discuss both common and unique issues related to developmentalism in the fields of early childhood education and middle grades education. In each chapter, an author from each field discusses critical issues regarding developmentalism.

In the first chapter, *A Schismatic Family and a Gated Community?*, Vagle and Parks describe and examine some of their assumptions about how each field has engaged in discussions surrounding developmental appropriateness and responsiveness. They begin by situating what it has meant to pay attention to development in early childhood and middle grades education, followed by a tracing of how the fields got to where they are today. They close by sharing some considerations for these fields as they move forward.

Based on their initial analysis of the fields, Vagle and Parks felt that the metaphors of a schismatic family (early childhood education) and a gated community (middle grades education) accurately portrayed the state of each field with regard to developmentalism. That is, in early childhood education the substantive conversations between developmentalists and reconceptualists appeared to have been divisive at times and that in middle grades education there did not appear to be such a divide because the conversation was not actually taking place. However, after analyzing seminal texts in each field (National Association for the Education of Young Children's position statement *Developmentally Appropriate Practice* and National Middle School Association's position paper *This We Believe*) and having informal conversations with leaders in both fields, Vagle and Parks came to see the schism and the gate not quite as sharply—seeing more *critical* promise than they had originally.

In the last chapter, *Walking the Borderland*, Conklin and Lee explore some key questions (i.e., "What do we mean when we say that we need to move beyond developmentalism?") that might be worth pursuing when scholars with differing orientations try to engage in honest and yet respectful conversations about their views of developmentalism within and across fields.

Reviewing some seminal works (e.g., Jackson & Davis, 2000; NAEYC, 1987, 1997, 2009) in the fields of early childhood and middle grades education, Conklin and Lee point out that developmentalists and critical theorists, albeit seemingly contradictory to one another in terms of their theoretical orientations, have had shared commitments to serving the educational interests and needs of young people at particular moments in their lives—whether those young people are categorized as young children or as young adolescents. These authors argue that acknowledging these shared commitments might be the first step toward listening across differences. Conklin and Lee identify the issue—how to characterize these young people without essentializing them—as a common task that scholars with differing orientations might want to grapple with together. They explore Parker's idea (forthcoming) of "talking to strangers" and Moss's notion (2007) of "agonistic pluralism" (p. 234) to gain insights into how different perspectives can be welcomed and appreciated in order to enrich our understandings of and work with young people at particular moments in their lives.

Clearly, the ultimate goal of this book is to critically examine the dominant discourse of developmentalism in the fields of early childhood and middle grades education. This goal may well be achieved in part by "taking a side" and by focusing only on the critiques of developmentalism. As mentioned in the general introduction at the beginning of the book, however, we are also concerned about the divisions between scholars with differing orientations in the fields. We believe the fields can continuously be reimagined when we include multiple perspectives that can be considered alongside one another. We hope that the readers can see our commitment to multiple perspectives through the two chapters presented in this last section.

References

Jackson, A., & Davis, G. (2000). *Turning points 2000: Educating adolescents in the 21st century*. New York: Teachers College Press.

Moss, P. (2007). Meetings across the paradigmatic divide. *Educational Philosophy and Theory, 39*, 229–245.

National Association for the Education of Young Children. (1987). S. Bredekamp (ed.), *Developmentally appropriate practice in early childhood programs serving children birth through age 8*. Washington, DC: Author.

———. (1997). *Developmentally appropriate practice in early childhood programs* (revised ed.; S. Bredekamp & C. Copple, eds). Washington, DC: Author.

———. (2009). *Developmentally appropriate practice in early childhood programs serving children birth through age 8* (3rd ed.; C. Copple & S. Bredekamp, eds). Washington, DC: Author.

Parker, W. C. (Forthcoming). Listening to strangers: Classroom discussion and political education. In S. Haroutunian-Gordon & L. Waks (guest eds), "Listening in Context," special issue of *Teachers College Record*.

CHAPTER 11

A Schismatic Family and a Gated Community?

Mark D. Vagle and Amy Noelle Parks

As described in this book's general introduction, there have been substantive conversations between developmentalists and reconceptualists in early childhood education for two decades (e.g., Lubeck, 1994, 1998; Mallory & New, 1994). This conversation continues today in venues such as the two early childhood special interest groups of the American Educational Research Association. Although such a debate is present in other fields dedicated to adolescents (e.g., adolescent literacy), it has been notably absent in mainstream middle grades education scholarship (Brown & Saltman, 2005). Instead, critical perspectives such as those published in a special issue of *Theory into Practice* dedicated to rethinking middle grades (e.g., Gay, 1994; Lesko, 1994) have remained on the periphery.

We came to see our respective fields this way through different paths that converged during several of our conversations. Amy's path to the field of early childhood education was forged as she transitioned to becoming an assistant professor. This transition involved moving away from thinking of herself as a mathematics educator concerned with young children to seeing herself as an early childhood educator concerned with mathematics. As part of this re-identification, Amy spent a significant amount of time in her first year tracing the history of early childhood education, both through written texts and through the stories of colleagues. In doing this, she was struck by two things: the centrality of the concept of *developmentally appropriate* to the field and the ways that this concept seemed to draw lines between colleagues in a variety of ways, including the conferences they attended, the journals they published in, and the ways they thought about their work in relation to early childhood education.

Mark's path involved less re-identification and more critique of what he had assumed about middle grades education. As a former middle grades teacher and administrator, Mark was well versed in the centrality of *developmental responsiveness* in middle grades education and had, without question, adopted a stance in his practice that reinforced this centrality. While pursuing his PhD, however, Mark began studying the work of critical theorists, phenomenologists, and post-structuralists. He began to question some of his own and the field's assumptions about young adolescent growth and change. As a new assistant professor, Mark's questions grew to become significant parts of his work, as he turned to some writing outside of mainstream middle grades education, namely that of Lesko, Brown, and Saltman, to inform his research and teaching. Mark began to see and believe that a dearly held conception such as developmental responsiveness could be read as problematic.

In trying to make sense of what we were experiencing, we had many discussions in which we found ourselves struck by the similarity of conversations in both middle grades and early childhood education, as well as the seeming lack of communication between these two fields that both define themselves primarily by the age of children with which they are concerned. These conversations convinced us that it would be productive for early childhood and middle grades educators to share the ways the two fields have defined and challenged concepts related to development and developmentalism, to examine the ways the conversation has been conducted in the two fields, and to explore some of the questions and challenges that both fields might productively take up in future work.

As the title of this chapter indicates, we came to this work with particular assumptions about how each field has engaged in discussions surrounding developmental appropriateness and responsiveness. Based on our initial analysis of the fields, we felt that the metaphors of a schismatic family (early childhood education) and a gated community (middle grades education) accurately portrayed the state of each field with regard to developmentalism. It appeared that in early childhood education the substantive conversations between developmentalists and reconceptualists had been divisive at times and that in middle grades education there did not appear to be such a divide because the conversation was not actually taking place. However, after analyzing seminal texts and having informal conversations with leaders in both fields we do not see the schism and the gate quite as sharply—in fact we see more *critical* promise now than we had originally.

With this in mind, we intend for this chapter to at once be a tracing of where these fields have been, an opportunity to find connections across the fields, and an effort to look forward toward critical ends. We begin by situating what it has meant to pay attention to development in early childhood and middle grades education, followed by a tracing of how we got to where we are today. We close by sharing some considerations as these fields move forward.

SITUATING DEVELOPMENT IN EARLY CHILDHOOD AND MIDDLE GRADES EDUCATION

To begin, we look briefly at how each field has mobilized the concept of development, discussing both similarities and differences. In early childhood, "developmentally appropriate practice" (commonly referred to as DAP) is the most frequently deployed phrase in conversations around development. The National Association for the Education of Young Children (NAEYC) has published three handbooks (NAEYC, 1987, 1997, 2009) defining and describing developmentally appropriate practice for early childhood educators. This document—sometimes referred to as "the green bible" by early childhood practitioners—has in many ways become the gold standard for what preschool and primary education is supposed to look like. In addition, the handbook is frequently cited by government and nonprofit organizations concerned with early childhood (e.g., Massachusetts Department of Education, 2003; National Research Council, 2001). As a result, this text occupies an important place at the center of the field of early childhood education in defining both what the field is and the issues with which it is concerned.

The 2009 document *Developmentally Appropriate Practice in Early Childhood Programs Serving Children Birth through Age 8* (NAEYC, 2009, pp. 9–10) describes DAP as including a knowledge of "age-related characteristics that permits general predictions about what experiences are likely to best promote children's learning and development," a concern for "individual variation," and a familiarity with the "social and cultural contexts in which children live." In this way, the authors of the handbook have sought to include responsiveness to individual and cultural variations within the umbrella of "developmentally appropriate."

Much of the rest of the text is aimed at providing examples and non-examples of developmentally appropriate practice for various age groups of children, including infants and toddlers, preschoolers, kindergarteners, and primary-grade (one–three) children. For example, teachers in a developmentally appropriate preschool classroom are described as engaging "in conversations with both individual children and small groups," and are contrasted with teachers who speak to children primarily with "brief comments or directions" (NAEYC, 2009, p. 165). Examples and non-examples are given across a wide variety of domains, including social relations, language, mathematics, and gross motor skills. In addition, this current version of the handbook includes a DVD with video examples of developmentally appropriate practice for use with preservice and in-service teachers.

Middle grades educators typically do not use the phrase *developmentally appropriate*, as much as they do *developmentally responsive*. National Middle School Association (NMSA)—the leading organization dedicated solely to the education of young adolescents—has published three editions of its position paper *This We Believe* (*TWB*; NMSA, 1982, 1995, 2003), and is releasing a fourth (forthcoming, 2010). *This We Believe* serves as the organization's vision statement for effective middle schooling and is considered an authority on middle grades education. In fact, the preface of the 2003 edition states that it is the "most widely used document on middle level education ever published. Schools have employed it as criteria for school evaluations, self-studies, parent and public education initiatives, and future planning. A series of position papers, research and curriculum summaries, and many research studies have been based on or derived from it" (p. x). Although *TWB* has not gained biblical status, it can be argued that it serves as the most clear and recognizable statement of what middle grades schooling should look and be like.

The titles of each edition vary in their direct reference to developmental responsiveness: the first edition entitled *This We Believe (TWB)*, the second *TWB—Developmentally Responsive Middle Level Schools*, and the third *TWB—Successful Schooling for Young Adolescents*; yet they all are based on developmentalism. Although not present in the title, this third edition embeds developmental responsiveness throughout the text. For example, with regard to teaching practices the 2003 edition states, "Developmentally responsive instructional practices place students at the center of the learning process. In such situations students are viewed as actors rather than audience" (p. 15). Similarly, curricular matters are described with developmental responsiveness as its referent. "In developmentally responsive middle level schools, however, curriculum embraces every planned aspect of a school's educational program" (p. 19).

Similar to NAEYC's DAP handbooks, the three editions of *TWB* demonstrate a commitment to individual variation and contextual factors, yet preserve the larger, foundational commitment to developmental responsiveness. For instance, in the 1995 edition NMSA tries to capture large contextual changes. However, there still is a retreat to developmentalism—"Guidelines for selecting educational goals, curriculum content, and instructional processes grow out of an awareness of this distinctive developmental age group" (p. 10)—and some "tinkering" with context—"When coupled with an equally full understanding of the cultural context in which youth grow to maturity, educators have the essential foundation for making wise decisions about educational programs" (p. 10). This same section of the 2003 edition reads more concisely, yet sends a similar message.

> The guidelines for selecting educational goals, curriculum content, and instructional processes grow out of an awareness of and respect for the nature of these distinct young adolescents. Educators who understand them and the cultural context in which they grow to maturity will make wise decisions about the kinds of schools needed. (P. 6)

The 2003 edition, much like the 1995 one, opens with a rationale for middle grades education and then describes the essential characteristics for successful middle grades schools. Unlike the DAP handbooks, *TWB* does not get to the level of specific practices, rather it describes to what ends curriculum, instruction, assessment, and so on should be aimed. Nevertheless, the DAP handbook and *TWB* are clearly central to the two fields in that they are widely distributed, are generally recognizable to practitioners and researchers who work in and/or identify in some way with the respective fields, and serve as guiding documents for policy, research, and practice.

Getting to Today

In this section we trace the conversations around developmentalism in both early childhood and middle grades education over the last few decades. In it, we argue that the notion of developmentalism shaped each field in different ways. These differences can help scholars in each field see their own histories around developmental conversations as less inevitable. We arrived at our assertions by analyzing the three editions of NAEYC's *Developmentally Appropriate Practice in Early Childhood Programs Serving Children in Programs from Birth to Age 8* (NAEYC 1987,1997, 2009) and NMSA's *This We Believe* (NMSA 1982, 1995, 2003). Our primary analytic question was: *How is developmentalism conceptualized in the texts and how does this conceptualization change over time?* Additionally, we had informal conversations with leaders in each field and draw on these conversations throughout our discussion.

Early Childhood Education—A Schism?

Much of the current conversation in early childhood education has been shaped by the first handbook on developmentally appropriate practice (NAEYC, 1987) and the critical response to it (e.g., Kessler, 1991a, b; Mallory & New, 1994). This first edition of the handbook sought to define developmentally appropriate practice for young children in response to the "increased emphasis on formal instruction in academic skills" (NAEYC, 1987, p. 1) in early childhood classrooms, which grew out of many politicians' critiques of public schooling. The editors of the handbook sought to intervene and redirect early childhood practitioners toward practices they saw as more humane and beneficial for young children. The handbook argued that developmentally appropriate practice was based on attention to "age appropriateness" as well as "individual appropriateness." In relation to development, the handbook argued that "human development research indicates that there are universal, predictable sequences of growth and change that occur in children during the first 9 years of life" (p. 2). For various age groups, the handbook noted expected developmental milestones in relation to cognitive, motor, and emotional development and offered lists of "appropriate" as well as "inappropriate" practices for each age group. For example, allowing five- to eight-year-olds to work cooperatively in small groups or alone in learning centers was described as appropriate, while requiring children to work silently and alone on seatwork was described as inappropriate, particularly when children were penalized for talking.

Although the 1987 handbook was disseminated widely by NAEYC and embraced by many practitioners, many early childhood educators in the academic community responded with vigorous critiques, both in writing and in person at various conferences (New & Mallory, 1994). In their edited book *Diversity and Developmentally Appropriate Practices*, Mallory and New (1994) identified many of the key thrusts of the critique, including the dichotomization of appropriate and inappropriate practices, the overreliance on Piagetian theory and "universal" developmental stages, the positioning of development as central to early childhood education, and the lack of attention to ethnic and cultural diversity.

Many of these same critiques were raised in other venues in response to both the first and second editions of the handbook. For example, Lubeck (1998) argued that given the increasing cultural diversity in the United States, it made little sense to work toward a shared vision of what practices all early childhood educators should adopt. She argued that in pushing toward a shared vision, the guidelines reduced possibilities for classroom practices in ways that would inevitably leave some children out. Fowell and Lawton (1992) suggested that the handbook diminished the role that direct instruction from the teacher could play in the learning of young children and criticized the book's tone, which they said promoted the idea that there was only one "right way" (p. 71) to think about early childhood teaching practices. Kessler (1991b) raised similar issues in her critique and suggested that the criteria of "democratic" replace developmentally appropriate as a way of thinking about productive teaching practices for young children. In doing this, she sought to undermine the psychological perspectives pervasive in child development conversations and to draw on critical theory to highlight goals for education other than individual progress toward developmental milestones. These goals might include helping children to think critically, to act in democratic ways, and to engage in social action.

One result of these varied critiques was the emergence of a divide in the early childhood community between members who relied primarily on the vision of developmentally appropriate instruction presented in the handbook and those who adopted critical perspectives, such as those described earlier. This divide showed up in a number of ways, including the creation of a yearly conference, *Reconceptualizing Early Childhood Education*; the launch of a new special interest group at the Annual Meeting of the American Educational Association—Critical Perspectives in Early Childhood Education; and the start-up of new journals, such as *Contemporary Issues in Early Childhood*. Each of these new venues for scholarly conversation, which were all launched in the 1990s, explicitly sought to encourage perspectives on early childhood that did not grow out of the field of psychology and concern for development. Rather than drawing primarily on the field of child development, these new organizations sought out research from a variety of perspectives, including post-structuralism, postmodernism, and sociology. For example, the stated purpose of the Critical Perspectives special interest group, which was created in 1998, is to encourage "the development of alternative perspectives and curriculum in Early Childhood Education" (AERA, 2009). From many of these alternative perspectives, research is not necessarily seen as empirical, but can be conceptual or philosophical in nature. In contrast, many of the longer-standing institutions in early childhood, such as the Early Childhood and Development SIG, NAEYC, and journals like *Early Childhood Research Quarterly*, emphasize practical significance, empirical research, and measurement and descriptions of child development.

While writing this chapter, Amy had informal conversations with a number of early childhood educators who located themselves differently in the field. Several of those associated with the newer SIGs, conferences, and journals spoke about feeling like "outsiders." For instance, Janette Habashi described the Reconceptualizing Early Childhood conference as "not mainstream" and said she felt there was a greater openness to diverse research perspectives there than in longer-standing early childhood conferences. Initially, Amy's reading and conversations led her to describe the field—as the title states—as a great schism. She felt that as a new professor she needed to choose a team and join up quickly.

However, after digging a bit deeper, Amy found that the divide was perhaps not as sharp and clear as it initially seemed. For example, many early childhood educators are members of both the newer and the long-standing SIGs. In fact, these two groups host a joint reception each year at AERA. In addition, many researchers publish in both the longer-standing journals and the newer ones. Finally, there have been several opportunities both in print and in person for early childhood researchers from differing perspectives to have conversations with each other. For example, Sue Bredekamp, one of the editors of the first NAEYC handbook, attended the Reconceptualizing Early Childhood Education conference in 1992, a time when many of the conference presentations were aimed at criticizing developmentally appropriate practice as described in the handbook. In a recent email exchange, Sue Bredekamp wrote that many of these early critical conversations were quite heated, but said that she thought the field was stronger for their intensity. She wrote:

> They strengthened the field by "maturing" it—made us think more critically about some of our long-held beliefs, avoid simplistic, knee-jerk reactions/answers to complex questions, take broader perspectives especially in the areas of culture and linguistic diversity, see the teacher's role as more complex and less passive, recognize that there are very few "universals," address the issue of whether and how these practices apply to children with disabilities—the list goes on and on. (S. Bredekamp, personal communication, May 13, 2009)

This perspective was mentioned by others, such as Linda Espinosa, who wrote in an email that "the open conversation and critique in

ECE has led to some really productive outcomes (particularly the new handbook)" (L. Espinsoa, personal communication, May 7, 2009). Indeed, the language used to describe developmentally appropriate practice in the 2009 handbook differs in many ways from the earlier versions. For example, in its first description of developmentally appropriate practice, the current handbook adds concern for children's "social and cultural contexts" (NAEYC, 2009, p. 10) to concern for developmental stages and individual differences. Also, rather than describe practices as appropriate and inappropriate the new handbook calls practices developmentally appropriate and juxtaposes these with other practices under the heading "in contrast." The practices under the in contrast heading are clearly portrayed as undesirable through descriptions such as "the kinds of things that well-intentioned adults might do but that are not likely to serve children well" (p. 75). However, the authors do note that "sometimes context affects whether a practice should be used or adapted" (ibid.). Throughout the book, the language about development, while still certainly infusing the message, is more tentative than in previous versions. Rather than talking about "universal, predictable sequences of growth and change" (NAEYC, 1987, p. 2), the new handbook talks about child development making "general predictions" possible (NAEYC, 2009, p. 9).

It seems likely that these changes, as well as the growing interest in culture, language, ethnicity, and dis/ability in many early childhood research settings, sprang from the critiques by both individual researchers and also research groups, such as the new SIG and journals. Clearly, the field is not so divided by a developmental schism that communication and critique does not occur. Perhaps, it is more accurate to describe the early childhood as a parliamentary legislature where members spend time caucusing with those who share their beliefs as well as time conversing with the larger body. However, the image of the schism may still be in some ways valuable. The notion that the critique has been productive in changing the language and the beliefs of the field may be more prevalent among those who dominate the mainstream perspective and have been in the position to choose which critiques will be incorporated and which will be left unaddressed. One early childhood colleague associated with the newer organizations speculated that the creation of special places for critical perspectives allowed researchers a place to speak without requiring the field to seriously change. She added: "There's so few writing and reconceptualizing the whole field. And their voices are not always heard."

Certainly, for example, the new handbook was not as significantly revised as many with more critical perspectives would like. The focus is still on developmentalism and the way culture is included often takes for granted white, middle-class norms and the need to "help" other children meet those norms, rather than questioning the power relations that maintain those norms in our schools and society. Children from low-income families are routinely described in terms of what they lack—for example, "they hear far fewer words" and "have dramatically less rich experience with language" (p. 2). These deficit-oriented descriptions are more common than ones based on children's strengths, which are currently unrecognized in most schools. In addition, although teachers are encouraged to acknowledge the cultures of families, at the core, the assumption is still that the teachers' job is to help parents change rather than to learn from parents and to change their own practices. For example, in discussing relationships with families, the handbook notes that the idea of collaborating with teachers is quite strange and uncomfortable for people in some cultures. The handbook recommends that "teachers need to be patient and let the relationship evolve gradually into a more collaborative, two-way partnership" (p. 183). The assumption that teachers ought to nudge parents toward the kind of relationship that they expect is one that many researchers with critical perspectives might identify as worthy of critique. While the current handbook and the current messages coming out of mainstream organizations such as NAEYC have changed, clearly the field of early childhood has not fully incorporated the perspectives of those who have sought to challenge mainstream views around developmentalism.

The challenge for researchers working in early childhood may be figuring out the extent to which the diverse research perspectives in the field should be integrated in the most important publications and conferences versus the extent to which researchers with differing perspectives ought to strike out and perhaps develop new fields or join other existing ones. One example of this move to separate is occurring in the field of special education, where researchers interested in cultural, critical, and postmodern perspectives have essentially created the new field of disability studies as a way of separating their work from the more traditional psychology-based work of mainstream special education. In some ways this separation might allow diverse early childhood researchers to more vigorously pursue their own interests. At the same time, if mainstream researchers believe the field has been strengthened by the critique and if researchers operating from alternative perspectives wish to have some impact on the people in the field most closely working with schools, children, and teachers, then breaking away may not be the most productive option.

Middle Grades Education—What Gate?

While writing this chapter, Mark learned that it is necessary to try to get a clear sense of what constitutes a "field." This was most apparent in Mark's informal conversations with Gayle Andrews (formerly Davis), coauthor of *Turning Points 2000* (Jackson & Davis, 2000) another seminal text in the field and David Virtue, current editor of *Middle School Journal*, NMSA's widely read practitioner journal. When asked about whether critical perspectives would be welcomed without creating a schism between developmentalists and critical theorists in the field of middle grades education, Andrews suggested that critical perspectives already are part of the scholarship if one defines the field as all who are writing about the experiences of young adolescents in some way (e.g., Brown & Saltman, 2005; Lesko, 2001). Virtue was mindful that middle grades scholars most often identify with middle grades education and at least one other field. For example, he also identifies as a social studies education scholar who has a strong interest and background in comparative education.

Andrews's and Virtue's perspectives gave Mark necessary pause to clarify what he thinks constitutes the field of middle grades education. Mark has, without hesitation, thought of mainstream middle grades education as those researchers, policymakers, and practitioners who actively advocate for young adolescents and identify in some way with organizations committed to these same efforts—namely NMSA and the Middle Level Research SIG of AERA. There are certainly other organizations interested in the education of young adolescents (e.g., the National Forum to Accelerate Middle Grades Reform—of which Andrews is the incoming president) and whose work constitutes the field of middle grades education. Also, the fields of mathematics, literacy, social studies, and science education, to name a few, also have researchers, policymakers, and practitioners who focus on the education of young adolescents in their respective content areas.

All of this said, for the purposes of this chapter Mark conceptualizes the field of middle grades education as any work associated with NMSA as it the most recognizable organization solely committed to the education of young adolescents. Therefore, the tracing Mark does here centers on the three editions of NMSA's position paper, *This We Believe*.

In 1992, NMSA reissued the 1982 edition of *TWB*, adding a foreword by the then executive director Denis Smith and a series of resolutions made by the Association in the ten years following the original *TWB*. The position statement itself remained the same. Mark cites the 1992 reissue when referring to the original (1982) edition.

In reflecting on the value of the 1982 edition, Smith stresses that *TWB* is based on a simple premise: "If we know the nature of the learner, then we must provide a school program to match that nature and thus nurture the adolescent that will emerge from the metamorphic state of human development variously called transescence, early adolescence, or prebuscence" (NMSA 1992, p. v).

From the start, there is an assumption that the learner has a "nature," that it can be known, and that schools must then respond to what is known. Moreover, the changes during young adolescence are amplified through the use of the word *metamorphic*, thus drawing special attention to this developmental stage. Although *developmentally responsive* is not as present in the 1982 editions as it is in the more current editions, it is foregrounded in some of NMSA's resolutions that followed. Resolutions during these ten years dealt with issues related to, for example, preparation of middle grades teachers, flexible grouping of students, interdisciplinary teaching team organization, and developmentally responsive instructional practices. One of the resolutions regarding teaching practices speaks to the importance of developmental responsiveness.

> 88–6. **Whereas**, the middle school learner is uniquely different from the elementary school or high school learner, and **Whereas**, when instructional methodology fails to respond to the unique qualities of the learner, student attitudes, achievement, and teacher efficacy can suffer. **Therefore, be it resolved that NMSA promote instructional practices which are developmentally responsive to the special characteristics of the young adolescent learner.** (P. 32; emphasis in the original)

The resolutions also contain a consistent emphasis on the uniqueness of this age group, sometimes through strong language such as "early adolescents differ in important ways from both younger children and older adolescents and others, and . . . are undergoing significant transitions, such as . . . '*perpetual metamorphosis*' and '*personality disorganization*'" (p. 30; emphasis added). The use of phrases such as these establish the urgency of this time in life and for responding to it in equally urgent ways. This urgency is also present in NMSA's search for a term that might accurately portray this unique time in life. There is concern that combining "an adjective and a noun as in young adolescent, early adolescent, late childhood, or emerging adolescent" (p. 5) would somehow diminish this urgency. They argued that the English language did not have a word for this time in life until Donald Eichorn (one of the committee members commissioned by NMSA to write *TWB*), in 1966, labeled it *transescence*, which according to Eichorn was

> the stage of development which begins prior to the onset of puberty and extends through the early stages of adolescence. Since puberty does not occur for all precisely at the same chronological age in human development, the transescent designation is based upon the many physical, social, emotional, and intellectual changes that appear prior to the puberty cycle to the time in which the body gains a practical degree of stabilization over these complex pubescent changes. (P. 5)

Clearly the use of an adjective and noun (young adolescence) has stood the test of time, as the 1995 and 2003 editions both refer to young adolescence rather than Eichorn's transescence. Developmental responsiveness also thrived, and perhaps reached its apex in the 1995 edition when it was added to the title—*This We Believe: Developmentally Responsive Middle Level Schools.* In the first paragraph one sees the centrality of developmental responsiveness. "In order to be developmentally responsive, middle level schools must be grounded in the diverse characteristics and needs of these young people" (p. 5).

The 1995 edition also includes a *Changing Society* rationale, focusing specifically on how changes during young adolescence are exasperated due to constantly changing societal contexts. However, NMSA reifies developmentalism when they assert, "cultures are evolving rapidly, and virtually every aspect of life has changed except for our children's *innate* developmental needs. Although children may mature physically more rapidly today, they still confront the same developmental hurdles as did previous generations" (NMSA, 1995, p. 8; emphasis added). One might wonder how all this can change but young adolescent's growth and change remains untouched.

It is also concerning to see children referred to in seemingly collective and static terms, as if the meanings ascribed to their growth and change are all the same. Throughout all three editions, issues regarding diversity are discussed in progressively more expansive ways, but the foundation itself (developmentalism) is not discussed accordingly. One resolution following the 1982 edition speaks to issues of diversity but resolves "to more effectively address the needs of our culturally diverse populations, both students and educators" (p. 35). The resolution, although seemingly generous, reinforces an insider/outsider mentality.

In the 1995 and 2003 editions, there also seems to be a call to "yesteryear" by referencing society's move away from nuclear families, presumably containing a mother and father in a married relationship. Although there is no direct reference to an ever-growing divorce rate, there is direct reference to the "absent father." Again, these changes in context seem to be treated outside of the conversation of developmentalism. It is as though context is described more as something that changes "as" young adolescents progress through the developmental stage, rather than the contexts actually saturating any conception of growth and change—including developmentalism.

The 2003 edition reads much like the 1995 edition, with a few exceptions. First, again, developmental responsiveness is no longer present in the title, but is embedded throughout the descriptions of the characteristics that constitute successful schools for young adolescents (e.g., multiple learning and teaching approaches that respond to their [young adolescent's] diversity). Second, the 2003 edition contains an expanded *call to action* in which NMSA suggests specific ways teachers, principals, parents, policymakers, and teacher educators can advocate for young adolescents. Third, the *characteristics of young adolescents* section in the 2003 edition is said to be research-based.

This is a significant addition as none of the editions directly reference existing research to support the other claims—something that has been of particular concern to Micki M. Caskey, current editor of *Research in Middle Level Education Online* and chair of *NMSA's Research Advisory Board*.

> I think what's been unfortunate about *This We Believe* is that it does not articulate its theoretical base…and it does not include how the position is grounded in research. One of the tasks of the Research Advisory Board this summer…is to identify research and resources in support of the new *This We Believe*… (A similar volume, Research and Resources in Support of This We Believe was published in 2003)…I think what's most exciting about the new version of *This We Believe* is for the first time in history there will be a number of pages that will actually talk about research…we actually have to have a theoretical or research base…for advancing our positions…I think that has been the largest critique of *This We Believe*…Upon what base does this work stand on? It can't just be this is what we believe without a stated premise undergirding it…I think if we went back to the earliest version that came out in the 80s, we would see a grounding in progressivism. But, the authors didn't actually write about that. (M. Caskey, personal communication, July 29, 2009)

Indeed, none of the existing editions contain citations and hence no reference section. Cakey attributes this, in part at least, to a concern (of NMSA's) that the text would not speak to practitioners in the same way. This prompts broader questions regarding the aims of *TWB*. If the first concern is about identifying the theoretical/research base that has undergirded such an important document for nearly thirty years, how does one get to the point (any time soon) in deliberating about how the field might reimagine this base—or how multiple theoretical perspectives might be used to inform *TWB*.

In middle grades education, perhaps one of the first steps to take is to introduce critical perspectives into the conversation and see what happens. Andrews, Caskey, and Virtue all felt this was possible and in Caskey's words "necessary." Andrews feels as strongly and advocates for a deliberate and purposeful effort to infuse critical perspectives into the conversation.

> We need to solicit, go after, recruit, make a strong campaign...to bring into the conversation, to bring into the publications, to bring into the presentations those folks who represent a different perspective...I do want the conversation to happen...it needs to happen...for lots of reasons...it seems as though we're operating in different circles. (G. Andrews, personal communication, July 22, 2009)

Caskey adds that the conversation should begin at the scholarly level more than at the practitioner level. Although I fully agree, this does pose an interesting dilemma. If the conversation begins and ends up residing primarily in scholarly conversations and does not make its way to practitioner conversations, it is not likely to influence documents such as *TWB* and as a result the field of middle grades education. Virtue stresses the importance of thinking about how critical theoretical perspectives will be received by "teachers, by principals who want information—real practical stuff they can use and less of the theoretical...And how to communicate some of these ideas in ways that will be useful, will empower them to actually do things in their classrooms and their schools" (D. Virtue, personal communication, May 21, 2009). Perhaps as with all complex endeavors, it requires a multifaceted, simultaneous approach. One approach that might be useful in middle grades education is one Amy has used in some of her research. The field of rhetoric offers tools—such as the consideration of audience, genre, discourse, and persuasion—that would allow researchers to consciously take up different perspectives depending on the work they want to get done, without necessarily asserting that certain ways of seeing the world were objectively "true." "By drawing on literary rather than scientific traditions, rhetoric offers researchers the opportunity to talk about the world in ways that cause others to think differently about social interactions without etching boundaries between people as the result of 'evidence' " (Parks, 2009, p. 16).

In this sense the field has an opportunity to infuse similar messages to multiple audiences (e.g., academic and practice-based) in multiple ways, rather than proceeding in a linear fashion.

Moving Forward

Early childhood and middle grades education are indeed in different places with regard to critical conversations around developmentalism and therefore should most likely proceed in different ways. That said, the fields can also learn from each other. For instance, scholars in middle grades education may, at first glance, want to avoid creating a schism. However, creating a tension among perspectives has not necessarily been, as Amy asserted, divisive. Such debates about foundational issues can be generative.

When asked to predict how the field of middle grades education might look five years from now if critical perspectives were infused into scholarly conversations in the field, Caskey states,

> We won't have moved very far, but I think what will happen...is that there will be conversation, there'll be dialogue, there'll be disagreement, there'll be debate, deliberation, all of those things that you want in an academic community. And I don't think that's really been, very much in the Middle Grades conversation. I think...we've all been saying the same thing, we're all sort of moving along in the same direction. (M. Caskey, personal communication, July 29, 2009)

Conversely, perhaps early childhood education could avoid the pitfalls of "gating" the developmentalist and reconceptualist caucuses. It is possible that each individual caucus could fall into the trap of moving along in the same direction and not communicating with one another and therefore reducing opportunities for scholars from different perspectives to learn from and challenge each other.

We also still wonder what both early childhood and middle grades scholars are really trying to say when they talk about developmentally appropriate or "developmentally responsive" practices. If we use "development" to describe everything that we see as good about pedagogy, we dilute the meaning of the word and bring a lot of baggage related to biology and psychology to our conceptions of good teaching—no matter how we situate or contextualize the development. Developmentalism is that powerful a discourse. As Kessler (1991b) argued, when we conceptualize children's learning as primarily a succession of predictable biological stages, we lose the opportunity to consider curriculum and pedagogy in relation to other concerns. For example, biological and psychological theories encourage us to think of poor and minority children's language development as lacking in relation to the predictable stages observed in white middle-class children.

In like fashion, Lesko (2001) lauds those who have tried to complicate developmentalism by saying that development differs by gender and across racial groups. However, she is not convinced that this work does enough disrupting because it leaves the larger commitment to unidirectional, linear growth and change intact. This is present in the 2003 edition of *TWB*. When race, poverty, and ethnicity are described as conditions that can influence development, a static norm is assumed and everything else is variation. Moves such as these must be interrogated so that race, class, and ethnicity are not continually viewed as conditions, but as ways we live in the world with one another.

Other kinds of theories and perspectives may encourage us to see broader political contexts, ways in which schools privilege some cultural practices but not others, and culture-specific ways of learning and growing. In the effort to open up the term developmentally appropriate, the new early childhood handbook includes knowledge of culture as part of the definition. However, it doesn't make a lot of sense to think about culture in terms of development. It may make more sense to think of knowledge of child or adolescent development as one domain of many that teachers must know in order to engage in powerful pedagogies rather than as an umbrella term that means all things good. In giving up the catch-all terms developmentally appropriate and developmentally responsive we are forced to become more specific and more articulate about the pedagogies we advocate. Do we mean practices that are humane, productive, democratic, creative, flexible—and intellectually, emotionally, and physically engaging? We also open the possibility of recognizing that sometimes these values will conflict with one another and acknowledge that teachers must sometimes make choices among them. If we see all good teaching as "developmentally appropriate or responsive," then it is intellectually difficult to consider the contradictions that inevitably arise in the complicated world of the classroom.

References

American Educational Research Association. (2009). *Special interest groups: SIG directory*. Retrieved September 24, 2009, from http://www.aera. net/SIGs/SigDirectory.aspx?menu_id=26&id=4714.

Brown, E., & Saltman, K. (Eds). (2005). *The critical middle school reader.* New York: Routledge.

Fowell, N., & Lawton, J. (1992). An alternative view of appropriate practice in early childhood education. *Early Childhood Research Quarterly, 7,* 53–73.

Gay, G. (1994). Coming of age ethnically: Teaching young adolescents of color. *Theory into Practice, 33*(3), 149–155.

Jackson, A., & Davis, G. (2000). *Turning points 2000: Educating adolescents in the 21st century.* New York: Teachers College Press.

Kessler, S. (1991a). Early childhood education as development: Critique of the metaphor. *Early Education and Development, 2*(2), 120–136.

———. (1991b). Alternative perspectives on early childhood education. *Early Childhood Research Quarterly, 6,* 183–197.

Lesko, N. (1994). Back to the future: Middle schools and the turning points report. *Theory into Practice, 33*(3), 143–148.

———. (2001). *Act your age: A cultural construction of adolescence.* New York: Routledge/Falmer.

Lubeck, S. (1994). The politics of developmentally appropriate practice: Exploring issues of culture, class, and curriculum. In Mallory & New (eds), *Diversity & developmentally appropriate practices* (pp. 17–43).

———. (1998). Is developmentally appropriate practice for everyone? *Childhood Education, 74*(5), 283–292.

Mallory, B. L., & New, R. S. (Eds). (1994). *Diversity & developmentally appropriate practices: Challenges for early childhood education.* New York: Teachers College Press.

Massachusetts Department of Education. (2003). *Guidelines for preschool learning experiences.* Malden, Mass: Author. Retrieved June 25, 2009. from http://www.eec.state.ma.us/docs/TAGuidelinesForPreschool LearningExperiences.pdf.

National Association for the Education of Young Children. (1987). S. Bredekamp (ed.), *Developmentally appropriate practice in early childhood programs serving children birth through age 8.* Washington, DC: Author.

———. (1997). S. Bredekamp & C. Copple (eds), *Developmentally appropriate practice in early childhood programs* (revised ed.). Washington, DC: Author.

———. (2009). C. Copple & S. Bredekamp (eds), *Developmentally appropriate practice in early childhood programs serving children birth through age 8* (3rd ed.). Washington, DC: Author.

National Middle School Association. (1982). *This we believe.* Westerville, OH: National Middle School Association.

———. (1992). *This we believe.* Westerville, OH: National Middle School Association.

———. (1995). *This we believe: Developmentally responsive middle level schools.* Westerville, OH: National Middle School Association.

———. (2003). *This we believe: Successful schools for young adolescents.* Westerville, OH: National Middle School Association.

National Research Council. (2001). *Eager to learn: Educating our preschoolers.* Committee on Early Childhood Pedagogy. Bowman, B. T., Donovan, M. S., & Burns, M. S. (Eds), Commission on Behavioral and Social Sciences and Education. Washington, DC: National Academy Press.

New, R. S., & Mallory, B. L. (1994). Introduction: The ethic of inclusion. In Mallory & New (eds), *Diversity & developmentally appropriate practices* (pp. 1–13).

Parks, A. N. (2009). Metaphors of hierarchy in mathematics education discourse: The narrow path. *Journal of Curriculum Studies.* First published online on September 7, 2009, at http://www.informaworld.com/smpp/content~db=all~content=a914582098.

CHAPTER 12

Walking the Borderland

Hilary G. Conklin and Kyunghwa Lee

> *Middle school developmentalists who have worked so hard for young adolescents and critical theorists who speak against injustice, oppression, and naïve history are not enemies. Both seek something better for young people, their schools, and the larger social world. So long as they remain out of touch with each other, both are diminished. Perhaps the first step toward an alliance is to show how the ideas of each enrich those of the other.*
>
> —Beane, 2005, p. xv

In our respective fields of early childhood and middle grades social studies education, the two of us often find ourselves sitting on the fence. We are drawn to the critical and postmodern perspectives that help us question the assumptions of universality and scientific truth embedded in the past century's grand developmental theories. We also appreciate the way these perspectives urge us to pay attention to the role of context and social construction in our understanding of young people. Yet at the same time, we find tremendous value in using particular notions of development to help us understand the unique qualities of young children and young adolescents. Although James Beane's quotation (2005) at the opening of this chapter speaks to the division between developmentalists and critical theorists in middle grades education, we find his observation also apt for early childhood education: both of our fields are diminished when we fail to take advantage of and learn from our varied perspectives.

We often wonder: What do we mean when we say that we need to move beyond developmentalism? Are we denying unique and important aspects of human development that people experience as they

move through different moments of their lives? What if the very constructs of early childhood and young adolescence that we critique
actually are essential to our fields? What if these constructs enable
educators and other stakeholders to recognize the importance of paying attention to learners at these particular moments of their lives?
Relying on developmentalism is limited, but wouldn't our neglect of
developmental perspectives be problematic, too? What do we as
educators really need to know to best serve our students?

In this chapter, we explore these kinds of questions and argue, like
Beane (2005) in the epigraph, for the need to create a space for debate,
dialogue, and careful listening among scholars exploring diverse
theoretical frameworks within and across the fields. We do so using
the image that titles our chapter: "walking the borderland."

When discussing this analogy at first, we realized that we each had
somewhat different images about walking the borderland. Kyunghwa,
for example, thought of the image of a lone and frightened traveler
who stays on a long borderline between two lands, a stranger to both
lands, who tries not to accidentally step in and upset residents in either
of these lands. It was the image of struggle and loneliness. Hilary's
vision for "walking the borderland," however, was more hopeful: she
thought of the image of opening ourselves to appreciate the different
views from where we stand on the fence and using the beauty within
each view to fuel us in our walk forward. Although her image was
more positive than Kyunghwa's, she, too, envisioned a solitary
journey—she was enjoying the views from both sides, navigating the
border with care, but she was walking alone.

Our different images may be reflective of different experiences we
each have had in our respective fields. For this reason, in what follows
we first describe each of our personal journeys with developmentalism
and discuss how the dilemmas we have encountered might provide
useful insights for other early childhood and middle grades educators. We then examine several seminal works in both fields to identify
shared commitments and challenges between developmentalists and
critical theorists. We argue that recognizing shared commitments
and goals might be the first step toward the journey of listening to
and respecting different perspectives.

Personal Journeys

Kyunghwa's Reflection

My college life in the Department of Early Childhood Education in
the mid- to late-1980s in Korea was filled with learning various stage

theories of human development. I remember having to memorize Jean Piaget's cognitive development stages, Lawrence Kohlberg's moral development stages, Sigmund Freud's psychosexual development stages, and Erik Erikson's psychosocial development stages in many courses. In particular, it was Piagetians' heyday, and lectures about various Piagetian experiments with children were common. In one course in particular my peers and I were extensively exposed to the curriculum developed by Kamii and DeVries, two well-known Piagetians, and were asked to design activities or game boards based on Piaget's theories. The notion of egocentric children who can't take another's perspective and who believe, for example, that the moon is following them wherever they go, promoted by Piaget's theory was fascinating. These grand theories led me to romantic perceptions of young children and to justify the importance of my work as an early childhood teacher, a profession neglected and degraded in many societies, including Korea. The ultimate goal of working with young children appeared to be helping these bright young scientists who are trapped in biological constraints make a smooth transition to the next stage of development.

While working as a kindergarten teacher, I constantly relied on the grand developmental stage theories to interpret my observations about children. When I came back to the university to pursue a master's degree in 1990, I learned that educators had begun using a new term, *developmentally appropriate practice (DAP),* as the first edition of the NAEYC's position statement (Bredekamp, 1987) had been translated into Korean and rapidly spread among Korean early childhood educators. Although this terminology was new, its basic premise based on Piaget's theory and maturationism (Walsh, 1991) was the same as what I had learned years ago. I took this familiar discourse for granted.

Four years later I found myself struggling in a graduate seminar titled "Post-Piagetian Perspectives and Implications for Early Schooling" that I took to pursue a PhD degree in early childhood education at a U.S. higher education institution. There I was first exposed to Vygotskian perspectives and various critiques about Piaget's theories. In addition to the language barrier that I experienced as a first year international graduate student, concepts and terminologies introduced in this seminar were so new to me that I had great diffi-culty in making sense of what I read and heard. I was disoriented.

Over the years, however, my life in two different parts of the world has made me convinced of the sociohistorical perspective and of how human development like all aspects of human life cannot be under-stood without considering cultural and historical contexts in which development is defined and occurs. I have been particularly drawn to a contemporary developmental perspective called cultural psychology that I discussed in my chapter of this book and other works (Lee, 2001; Lee & Walsh, 2001).

While exploring a contemporary developmental perspective with a keen interest in critical and postmodern perspectives, I have witnessed two highly charged debates among early childhood educators with different perspectives. One was during a symposium held in 1997 to honor a renowned early childhood educator's retirement. I saw audiences divided into two groups, one defending the notion of DAP promoted by NAEYC and the other, albeit a minority, challenging the assumed consensus in DAP. Members of each group cheered a speaker who represented their perspectives. The other debate, which occurred at the 2002 annual meeting of the American Educational Research Association (AERA), moved further beyond the legitimacy of DAP and contested whether child development knowledge was essential to early childhood teacher education. Although no audiences visibly cheered a speaker this time, extremely opposite perspectives were presented during the session.

I am a member of two early childhood special interest groups (SIGs) of the AERA: the Early Education and Child Development SIG and the Critical Perspectives on Early Childhood Education SIG. Although I have noticed that some sessions are organized collaboratively by the two SIGs, I still see the division between the two groups and feel pressure to choose and develop a clear alliance to one of them in order to feel a sense of belonging. Yet I can't help but think how both relying on the dominant developmentalism and rejecting insights about human development are equally limiting. I feel like I haven't found a comfortable place in the field yet.

Hilary's Reflection

My development as a researcher of middle grades teacher preparation began among many professors and fellow doctoral students who often questioned my characterizations of young adolescents. I recall with some emotion the mutual resistance that occurred when I made efforts to describe the middle school students who motivated my research. Drawing upon my classroom teaching experiences in sixth, seventh, and eighth grades, I talked to my doctoral classmates about young adolescents as funny, energetic, creative, quirky, full of questions, and brimming with intellectual potential. They were at a critical time in their schooling, ready and *needing* to be engaged.

Yet as I spoke in my doctoral classrooms of the exciting cognitive development that occurred in the age range of ten–fifteen years, many of my classmates would critique my claims to the uniqueness of the middle school years and dismiss my essentializing of young adolescents. Indeed, the very word "development" was a conversation-stopper among many in my curriculum and instruction department, because it signified the positivistic categorization of human lives that those others in the human development and educational psychology department were rumored to enjoy. I often wondered to myself in the midst of these discouraging conversations: Have *you* ever taught middle school students? Do *you* know what they're like? Do you care about them and want to advocate for them the way that I do? My classmates' critiques seemed purely intellectual, without grounding in experience or empirical understanding. I felt resentful of those who questioned my characterizations without sharing my commitments.

Amidst these seemingly unfriendly challenges to my research passions, I found great comfort in the existing research literature on young adolescents and the preparation of middle school teachers. These authors' descriptions of middle grades students resonated with my own teaching experience; they spoke to the rapid and exciting physical, social, intellectual, and emotional changes that I had witnessed among my own sixth, seventh, and eighth grade students. While I had studied general theories of adolescent development as an undergraduate and in my master's teaching certification program, this was the first time I had encountered research literature that specifically addressed *young* adolescents—a group of young people who I felt were neglected and warranted special attention. The authors I read during my doctoral years seemed to understand the common questions I had been asked when I told people I was a middle school teacher: "Are you an angel?" or "Are you crazy?" These authors' descriptions not only supported my own understanding of the nature of young adolescents, but clearly came from people who were, without a doubt, passionately committed to serving the needs of middle grades students and deeply concerned about the education of these young people, as I was.

Seven years later, I can see the conversations with my classmates with less defensiveness, and I can now better appreciate both my own and my classmates' perspectives. While I stand by my commitments to and my general characterizations of young adolescents, I now better understand how bringing attention to puberty can be potentially paralyzing to young people, how characterizing young adolescents' intellectual development can lead educators to minimize their own role in facilitating such development, and how describing middle school students' physical, social, emotional, and cognitive development often fails to honor the culture, race, social class, and gender that significantly shape young people's lives and experiences in schools.

At the same time, I still find myself engaging in "Yes, but—" conversations in which it feels like colleagues and I are talking past each other. Although I now readily acknowledge the socially constructed nature of early adolescence, I still try to persuade others that there is something unique about young adolescents. I find myself in conversations like this:

> *Hilary*: I think one thing we can say that is unique about young adolescence is the emerging sexuality that happens during this time period.
> *Colleague*: But you know, even infants and young children show signs that they are exploring their sexuality.
> *Hilary*: Yes, but, can't we agree that girls getting their periods and the sexual maturation that occurs during young adolescence is distinct from that exploration in earlier childhood?

In conversations such as these, I feel pegged to developmentalism once again, thrown from what had seemed like my comfortable border path, back onto one side of the fence. I find myself wondering: *Am* I clinging too much to biology? Isn't there some middle ground? I'm still searching for how I, along with others, can embrace the views on both sides of the fence.

Navigating the Essentializing Border

The Need to Name Young People

Building on our experiences, we now consider how we and others might go about navigating this borderland. Before discussing some shared commitments between developmentalists and critical theorists, we would like to note one issue. Oftentimes, critiques of the notion of human development stages and their assumed universality focus on developmental psychology as if this particular discipline created such a notion. Yet the idea of stages of human development predates the discipline of developmental psychology's emergence in the late nineteenth century. Cannella (1997), for example, traced the notion of stages as early as the "Greek belief in the number 7. The Greeks considered age 7 to be the age of reason, 14 to be adolescence, and 21 to represent maturity" (p. 55). Walsh (1991) also described the Protestant reformer Martin Luther's (1483–1546) idea about changes in human life in every seventh year. As Lee and Vagle discussed in the general introduction of this book, Jean-Jacques Rousseau (1712–1778) and Friedrich Froebel (1782–1852) proposed stages of children's development as well (Wolfe, 2000). Given this history and prevalence, we can reasonably argue that stages of human development are not a fact or a concept discovered by developmental psychologists, but a cultural belief reified by the rhetoric of science and various social realities (e.g., rites of passage).

Indeed, scholars have argued how childhood and adolescence are not mere descriptors of biological maturation, but are social and historical constructions (e.g., Corsaro, 1997; James et al., 1998; Saltman, 2005). In particular, Cannella (1997) argued that childhood "has been created as a universal truth from within enlightenment/ modernist perspectives that focus on scientific reason" (p. 42). Whether accepting as universal and biological stages of maturation or as sociohistorical constructs, however, we believe both developmentalists and critical theorists would agree that without the notions of early childhood and young adolescence, there would have been no fields of early childhood and middle grades education.

Indeed, both of our fields rely on our ability to name the groups of young people we wish to serve. Early childhood reconceptualists critique stage developmentalism and argue that the dichotomous construction of child separated from adult can lead to "an implicit form of subjugation" (Cannella, 1997, p. 36). Yet, these scholars still identify their work with a field recognized by the specific age group or the specific period of human life. In the case of middle grades education, scholars tend to be affiliated with at least one other field, usually content areas, besides middle grades education (see chapter eleven, this volume). However, although critical theorists might not identify themselves as mainstream middle grades educators, they, too, write about young adolescents. Thus, at least to some degree, developmentalists and critical theorists in the fields share a commitment to advocating for these young people at particular moments of their lives.

Reasons to Essentialize

One of the major differences between scholars with differing orientations might be whether and how one essentializes particular characteristics of young children or young adolescents to advocate for them. A wise professor once said to Hilary, "it's possible to characterize without essentializing" (S. Schweber, personal communication, 2006). This statement highlights some of the tensions and contradictions around identifying some of the salient qualities of particular age groups. Even the term "age groups" in the previous sentence creates tensions because, as Graue (chapter five, this volume) and Rogoff (2003) articulated, the notions of time and age themselves are arbitrary and cultural. Yet, naming particular groups of people as young children or young adolescents and grappling with descriptions about some attributes of these people can provide a common language or a common point of departure for educators.

Essentializing groups of young people has often served political purposes. For example, early childhood has been essentialized as the most malleable and the most critical stage of human development. This essentialization has allowed the field to draw attention from politicians, the general public, parents, and educators and to justify early education and intervention for the security and economy of the nation's future. This essentialization of early childhood in connection to national agendas can be traced back to Plato, who emphasized education for early years and argued in the *Republic* that all children should be "taken away from their parents at birth and reared by the state" (Braun & Edwards, 1972, p. 12) to help parents as the citizens of Athens focus on their responsibility for "the public good" (ibid.). He also thought that many parents were not good guardians for the future citizens of an ideal republic (Cannella, 1997). Although Plato's position might be considered an extreme, the Work Projects Administration (WPA) nurseries established to support adult employment during the Great Depression of the 1930s, the Lanham Act nurseries to help women's participation in the war effort during World War II, and the creation of Head Start for the War on Poverty in the 1960s clearly show relations between national interests and early childhood education in the United States (Nourot, 2005). Today, this tendency still permeates as repeatedly seen in a variety of political rhetoric for local, state, and national elections, including the recent U.S. presidential election. Reporting on then president-elect Barack Obama's pledge for a ten-billion-dollar investment in early childhood education, "the largest new federal initiative for young children since Head Start," a columnist in the *New York Times* wrote:

> Mr. Obama's platform accepts the broad logic of the Ypsilanti study. "For every one dollar invested in high-quality, comprehensive programs supporting children and families from birth," the platform says, "there is a $7–$10 return to society in decreased need for special education services, higher graduation and employment rates, less crime, less use of the public welfare system and better health." (Dillon, December 16, 2008)

This economic justification for the education of young children along with the logic of education for global competitiveness and better economy has been popular among politicians. In this rhetoric, we see the influence of Plato who argued: "[The] beginning is the most important part of any work, especially in the case of a young and tender thing; for that is the time at which the character is being formed and the desired impression is more readily taken" (as cited in Braun & Edwards, 1972, p. 13). Here, young children are essentialized as being flexible and easy to mold, characteristics well accepted by the general public and many educators. Coupled with these characteristics, J. McVicker Hunt's *Intelligence and Experience* (1961) and Benjamin Bloom's *Stability and Change in Human Characteristics* (1964) promoted the idea of the early years as the most important period for intellectual development.

Essentializing plasticity and intellectual potential in young children, however, creates unexpected tensions as the emphasis on early education often results in "trickle-down" academic pressure on young children. For example, although public kindergarten programs, which began in St. Louis public schools in 1873, and the movement for public school sponsored preschool programs since the 1990s have increased access to early childhood education for many children, educators have also been concerned about the accompanying "academic pushdowns"—a phenomenon that expects kindergarteners and preschoolers to learn and perform academic skills usually taught in primary grade classrooms (Nourot, 2005).

In addition, as discussed in the general introduction of this book, schooling practices in the United States have increasingly focused on high-stakes testing since the 1980s. The urgency of educational reforms based on standardized tests was promoted by the report *A Nation at Risk* published by President Reagan's National Commission on Excellence in Education in 1983 that detailed American students' poor academic performance in comparison to their counterparts in other industrialized countries and its possible negative impact on the nation's future economic status in the world (Shepard, 1999). The trend of test-based educational reforms and the emphasis on academic accountability have further intensified since the *No Child Left Behind (NCLB)* Act of 2001. Early childhood educators' concern about academic skills oriented curriculum and narrowly defined standards has increased accordingly (e.g., Stipek, 2006).

The publications of the series of editions of DAP by the NAEYC are reflective of early childhood developmentalists' concern about early schooling practices in recent years and the unexpected negative impact of the way young children's plasticity and intellectual potential has been essentialized. The first edition, for example, clearly stated that the publication of the DAP guidelines was to respond to the "increased emphasis on formal instruction in academic skills" (Bredekamp, 1987, p. 1) and described such instruction as developmentally inappropriate for young children. The second edition repeated this concern about "a growing trend toward more formal, academic instruction of young children—a trend characterized by downward escalation of public school curriculum" (Bredekamp & Copple, 1997, p. v). This revised version elaborated: "The primary position was that programs designed *for* young children be based on what is known *about* young children" (ibid.). Compared to these previous editions, the third edition emphasized continuity across early childhood and elementary education, particularly across grades preK-3 (Copple & Bredekamp, 2009). While acknowledging the concern about poor student achievement and "its impact on American economic competitiveness in an increasingly global economy" (p. 2) as well as the good intent of *NCLB* (e.g., making schools accountable for closing achievement gaps between children from different backgrounds), this most recent version of the DAP guidelines also revealed its concern about the negative impact of academic accountability on early childhood classrooms:

> At the same time, however, preschool educators have some fears about the prospect of the K-12 system absorbing or radically reshaping education for 3-, 4-, and 5-year-olds, especially at a time when pressures in public schooling are intense and often run counter to the needs of young children. Many early childhood educators are already quite concerned about the current climate of increased high-stakes testing adversely affecting children in grades K-3, and they fear extension of these effects to even younger children. (P. 4)

Unlike the political rhetoric that tends to essentialize young children as malleable human resources who need to be tamed and shaped as early as possible for the nation's future, early childhood developmentalists concerned about the watered-down academic curriculum essentialize young children as individuals who are qualitatively different from older children or adults and who need to be protected from excessive expectations and pressures. In particular, one of the NAEYC's commitment statements emphasized "appreciating childhood as a unique and valuable stage of the human life cycle [and valuing the quality of children's lives in the present, not just as preparation for the future]" (Bredekamp & Copple, 1997, p. 7). This statement revealed the field's long-held belief in and respect for childhood that could be traced back to what Rousseau wrote in *Emile*: "Hold childhood in reverence... Give nature time to work before you take over her business" (as cited in Wolfe, 2000, p. 40). We find it ironic that although the DAP guidelines emphasized valuing children's present lives and their childhood, in practice educators often provide activities under the name of DAP that are not meaningful for the children's present (and even future) lives as discussed by several chapters in the first section of this book.

Just as early childhood educators essentialize young children not only to draw national and political attention, but also to direct this attention to what they consider proper pedagogy for young children, middle grades educators, too, have often essentialized young adolescents in order to draw attention to and advocate for these neglected young people. Policymakers, funders, and others who hold power to make change seem to be more compelled by projects that address crises, critical needs, dire situations, and clearly defined communities. It is for these reasons, then, that middle level advocates often craft arguments such as this: "[Early] adolescence is a period of both enormous opportunities and enormous risks. Although many young people reach late adolescence healthy and ready for the challenges of high school and adult life, early adolescence for many others is the beginning of a downward spiral" (Jackson & Davis, 2000, p. 8). While these authors of *Turning Points 2000* could have framed early adolescence as simply a time when most young people experience the typical ups and downs of human life, readers are more likely to take note of strong language that clearly articulates a pressing concern about an underserved population. Indeed, Joan Lipsitz's seminal report on early adolescence, *Growing Up Forgotten* (1977), sounded an alarm about both the neglect of and the myths surrounding ten-to fifteen-year-olds. Her stark title is a reminder that the language we choose in advocating for young people matters. In this way, then, persuading the public that young adolescence is a critical turning point becomes a skillful political tool to help ensure that the young people we want to serve do not grow up forgotten.

At the same time, the same strong, rigid language that brings attention to the middle grades can be problematic. We need to simultaneously honor the value of the sometimes black and whiteness of documents like *Turning Points* and the purposes that this language serves, while also acknowledging that this stark language can lead to unintended consequences. Like the DAP guidelines over the years, authors of the *Turning Points* reports have both revised some of their earlier language as well as reinforced some common images of young adolescence. For example, Jackson and Davis (2000) explained:

> The original *Turning Points* report of 1989 drew attention to early adolescence as a fascinating period of rapid physical, intellectual, and social change. It is the time when young people experience puberty, when growth and development is more rapid than during any other developmental stage except that of infancy. Dramatic physical changes are accompanied by the capacity to have sexual relations and to reproduce. It is a time, too, of emotional peaks and valleys, "of trial and error, of vulnerability to emotional hurt and humiliation, of anxiety and uncertainty that are sources of unevenness of emotions and behaviors associated with the age" (CCAD, 1989, p.21). Yet within the trials and tribulations of early adolescence are the opportunities to forge one's own identity, to learn new social roles, and to develop a personal code of ethics to guide one's own behavior. (Pp. 6–7)

Although the language of the 2000 report places greater emphasis on the opportunities embedded within the changes of young adolescence rather than the risks, both reports reinforce the classic notion of the drama, storm, and stress of this time. And, it is these pervasive images of young adolescence, in turn, that often lead middle grades educators to believe that young adolescents are too preoccupied with emotional and hormonal tumult to be able to engage in substantive intellectual work (e.g., Conklin, 2008, chapter three in this volume; Lexmond, 2003; Yeager & Wilson, 1997).

Thus, it is here that we think developmentalists and critical theorists can help each other and learn from each other. We can use the strengths of these different perspectives to advocate for young children and young adolescents in strong but nuanced ways. We need to ask: How or when or under what circumstances is it ever helpful to essentialize young children and young adolescents? Or, alternatively, how can we characterize these groups of young people in ways that serve them *without* essentializing them?

CHALLENGES OF CHARACTERIZING WITHOUT ESSENTIALIZING

In fact, while being conscious of the condition and the way in which we essentialize young children and young adolescents, simultaneously we may also want to disrupt certain depictions of these young people. Carefully thinking about how those depictions limit people's perceptions about young children and young adolescents will help us identify some other characteristics that, ironically, we may deliberately want to essentialize in order to advocate for young people. We believe this might be a shared challenge and a topic for dialogues between developmentalists and critical theorists.

Among the essentialized characteristics of young children in the field of early childhood education, for example, the emphasis on young children's plasticity and on early intervention has often led to the characterization of young children as being passively molded by adults and the surrounding environment. This characterization is particularly problematic as it creates a deficit perspective of young children in general and of those from economically disadvantaged families and cultural and linguistic minority groups in particular—this perspective still exists in the most recent version of the DAP guidelines as discussed by Vagle and Parks in the previous chapter. In addition, the strong influence of Piaget's theory has essentialized young children as being smart but naïve as they are constrained by their biological stage. Coupled with this view, the field's dearly held belief in "natural unfolding," an idea based on maturationism, has characterized young children as constantly needing adult protection and guidance—the image of a plant in a garden as symbolized by Froebel's kindergarten (children's garden). These are characterizations that warrant challenging. Instead, we can characterize young children as active and intellectual meaning makers (Bruner & Haste, 1987; Donaldson, 1978) who have a sense of agency and resilience. We can conceptualize these children's development as a bidirectional process of shaping and being shaped by culture (Bruner, 1996; Corsaro, 1997; Shweder et al., 1998).

Similarly, among the characterizations of young adolescents in the field of middle grades education, we, like many critical theorists (e.g., Lesko, 2005), would like to disrupt the perception of young adolescents being hormonally driven and intellectually incapable, yet still mark them as distinctive, special, and worth paying attention to. If we are going to essentialize characteristics of young adolescents, it would seem to serve them better if we portrayed them as intellectually curious and capable cultural beings, without hormones and physiological characteristics used as their defining feature (Beane, 2005; Lexmond, 2003). Indeed, as Beane (2005) notes, "few middle school advocates seem to understand that by emphasizing development alone, the middle school concept fails to attach itself to any large and compelling social vision that might elevate its sense of purpose, attract more advocates, and help sustain the concept against its critics" (p. xiv). A cautious reconstruction of young adolescence has the potential to advance the cause of students in the middle grades.

However, this does not mean that young adolescents' physiological characteristics should be ignored. In fact, just as teachers should recognize the role that social class, race, and culture play in their students' lives, it can be helpful information for teachers to understand that some girls in their classrooms may be encountering the unfamiliar sensations of getting their periods, while some boys might be feeling self-conscious about being significantly shorter in height than their female classmates. Teachers can use this information to be understanding with their young adolescent students, while still emphasizing the intellectual work that they are in school to accomplish. As Saltman (2005) wisely clarifies:

> To claim that adolescence is a social construction is not to say, for example, that puberty is a fiction or merely a narrative with no natural scientific content. However, to recognize that adolescence is a social construction is to recognize...that the meaning of biological or psychological realities do become meaningful or relevant in different ways in different social contexts. (P. 16)

Thus, we need to recognize puberty, but not use it to define students or place limits on what we ask them to do.

Considering Multiple Perspectives and Nuances

We think our respective fields have not embraced diverse perspectives within and outside the fields, including contemporary developmental frameworks. In fact, both developmentalists who rely on the grand developmental perspectives and critical theorists who question the legitimacy of developmental theories tend to essentialize developmental psychology. As illustrated in chapter two in this volume, while the field of developmental psychology has begun to move beyond the grand developmental theories by the end of the past century (Damon, 1998), educators maintaining allegiance to these theories and those critiquing the theories characterize the field based on the outdated perspectives.

For example, early childhood reconceptualists have aptly pointed out how the first edition of the DAP guidelines (Bredekamp, 1987) emphasizing age appropriateness and individual appropriateness of curriculum and pedagogy for young children reflected the Western, particularly middle-class European American, and the individualistic notion of development (Cannella, 1997; Mallory & New, 1994). In response to this critique, the next two revised editions of the DAP guidelines acknowledged the changing demographics of children and families served in today's early childhood programs and included the importance of considering "the social and cultural contexts" for the decision-making for practice (Bredekamp & Copple, 1997; Copple & Bredekamp, 2009).

Yet, the social and cultural contexts were discussed in the revised guidelines as if they were an independent variable that could be separated from the individual child (the dependent variable). The 1997 edition of the DAP guidelines stated: "Rules of development are the same for all children, but social contexts shape children's development into different configurations" (Bowman, 1994, p. 220, as cited in Bredekamp & Copple, 1997, p. 12). According to this view, development is essentially a universal process and marked by cultural variations, a perspective critiqued by contemporary developmentalists (see chapter two in this volume). Although the 2009 edition of the DAP guidelines managed to avoid such an explicit claim, the basic assumption remained the same. As Vagle and Parks discussed in the previous chapter, this recent version of DAP constantly compared the achievement of children from low-income families or those from cultural and linguistic minority groups with that of their affluent, white peers in order to describe gaps between the children. Doing so, the guidelines revealed the notion of universal development affected by social and cultural variables. In this conceptualization, culture appears to determine development. This conception has often led to a deficit perspective of certain groups of children and to describe "children of color, children growing up in poverty, and English language learners" as lacking in academic skills and falling "farther behind with time" (Copple & Bredekamp, 2009, p. 6).

In addition, all three editions of the DAP guidelines (Bredekamp, 1987; Bredekamp & Copple, 1997; Copple & Bredekamp, 2009) considered age-related characteristics and individual variation as the important criteria for decision-making for educational practice. This idea, coupled with the aforementioned limited conceptualization of culture, attests to the field's reliance on Cartesian dichotomies of biology-culture and nature-nurture, which were prevalent in the grand developmental perspectives and critiqued by contemporary developmental theorists utilizing the systems models (Lerner, 2006), including cultural psychology (Bruner, 1996; Cole, 1996; Kitayama & Cohen, 2007; Shweder et al., 1998).

Yet as much as early childhood developmentalists have relied on outdated developmental theories, reconceptualists' critiques have also centered on the past century's grand developmental perspectives. While contemporary developmental theorists utilizing the systems approach have rejected "the positivist and reductionist notion" of universality in human development by embracing postmodern perspectives (Lerner, 2006) and explored "psychological pluralism" (Shweder et al., 1998), reconceptualists even in some recent critiques have not acknowledged these efforts and have continued to characterize developmental psychology as a field promoting "universal truths" and "the gendered and Western cultural assumptions" (Cannella, 2005).

One might attribute this tendency among reconceptualists to the fact that the dominant developmental knowledge has been the past century's grand developmental perspectives. In fact, Cannella (2005) and other reconceptualists (e.g., Ryan & Grieshaber, 2005) acknowledged how Vygotskian and sociocultural perspectives had influenced the dominant child development knowledge and reconceptualists' conception of " 'development' itself [as] a cultural (mainly western and male) construction" (Cannella, 2005, p. 28). Nonetheless, contemporary developmental perspectives have been omitted from alternative frameworks (e.g., postmodern, feminist, post-structural, postcolonial perspectives) that reconceptualists have been exploring. Although different philosophical orientations that have founded the disciplines of psychology and sociology might be a reason why reconceptualists are hesitant to incorporate contemporary developmental frameworks, Ryan and Grieshaber (2005) explained that the concern about "child development retaining its prominent position" (p. 35) in the field as another reason.

If our goal is to consider multiple perspectives to help our work with young people, we will need to include up-to-date frameworks from various disciplines in our exploration. Staying with a dogmatic and outdated perspective or avoiding a new framework for fear of its dominance will be equally short-sighted, particularly when the new idea has the great potential to help developmentalists and critical theorists come close to each other. We think our fields need to expand the repertoire of alternative perspectives.

In the same way, we also need to be alert to the fact that authors in our respective fields have sometimes positioned their arguments in more nuanced ways than their critics suggest, and that some critics are guilty of essentializing the diverse body of literature they are critiquing. For example, Lesko (2005) explains: "[Literature] on middle school practices so heavily emphasizes the physiological turmoil of young adolescents that self-esteem issues and hormones appear to consume them. Such an emphasis positions teachers to question whether such hormonally burdened young people can respond capably or successfully to substantive intellectual tasks" (p. 88). While we have already noted our agreement with the general sentiment Lesko is advocating here, the boldness of her claims about the middle school literature is not totally unlike the boldness of some of the literature she seeks to critique. With no references attached to her claims, we are left to think that all of the prominent literature in the middle school field must ascribe to these characterizations of young adolescents. Yet, as one example, the seminal *Turning Points 2000* report (Jackson & Davis, 2000) does not fully support Lesko's argument. While Jackson and Davis do, indeed, describe the physiological changes that typically occur among young adolescents, including the onset of puberty and rapid physical growth and change, they also strongly emphasize the intellectual focus of middle grades education. For example, they explain:

> What is the purpose of middle grades education?... Unfortunately, in the history of middle grades education, the purpose has at times become obscured. Let us be clear. The main purpose of middle grades education is to promote young adolescents' intellectual development. It is to enable every student to think creatively, to identify and solve meaningful problems, to communicate and work well with others, and to develop the base of factual knowledge and skills that is the essential foundation for these "higher order" capacities... middle grades schools must be about helping all students learn to use their minds well. (Pp. 10–11)

Thus, it seems that Jackson and Davis in fact share Lesko's interest (2005) in engaging middle grades students in substantive intellectual tasks—a message that is obscured by Lesko's bold claims. Scholars who are coming from these differing perspectives need to be cautious in their claims, because each side has the potential to alienate the other by not acknowledging what the other side contributes, and in doing so, both risk having less of an impact on the lives of young people.

Just as this disjuncture signals the need for careful reading and listening across our fields, we also need to take note of the distinctive perspectives within our fields. For example, George and Alexander (1993), two of the early leaders of the middle school movement, offer this careful description of the qualities of exemplary middle school teachers:

> The literature of middle school education is replete with references to the special characteristics desirable in persons electing to teach middle school youngsters... Yet we believe that the qualities that distinguish a good middle school teacher from a good elementary teacher or high school teacher are probably quite limited, albeit critically important.

> We believe that effective middle school instruction is implemented by
> teachers who have the flexibility of the generalist, the expertise of the
> specialist and the enthusiasm of one who understands and enjoys the
> special nature of the middle school student. (P. 142)

With this comment, George and Alexander distinguish themselves
from other middle level advocates (who, admittedly, they also do not
reference specifically), illustrating the need to read all authors' claims
carefully and not assume that all advocates in the field share a common set of perspectives or understandings.

Conclusions

As Beane (2005) at the opening quotation of this chapter suggested,
early childhood and middle grades developmentalists and critical
theorists "are not enemies." Nonetheless, the division between these
groups in our respective fields resembles in many ways the battle
between the developmentalists and the social meliorists, two of the
four interest groups who fought for dominance in the American curriculum during the first half of the twentieth century, a history well
documented by Herbert Kliebard (1995).

Using the analogy of walking the borderland, in this chapter we
have hoped to invite scholars with different perspectives and orientations to the effort to create a space for honest and yet respectful dialogues. To pursue such dialogues, we find Parker's notion
(forthcoming) of "talking to strangers"—his adaptation of Danielle
Allen's work—a potentially helpful way of thinking about how we
might continue such conversations. While Parker is writing about
classroom discussion as a tool for political socialization, his focus on
bringing together diverse people who may not know or like each
other to pursue shared problems seems apt for the discussion we present here. Drawing on Allen's work, Parker emphasizes that both our
understanding of common problems and our pursuit of what to do
about them *requires* diverse people coming together because it both
allows a shared, democratic decision-making process and gives us
much wider access to alternative interpretations and solutions than we
are able to generate in isolation.

In order to pursue such conversations, Parker argues that we must
learn to listen across difference. He explains, "Equitable and trust-
worthy conjoint living is not only a matter of being heard but also of
hearing others" (p. 11). In order to help us be heard and hear others,
he proposes a set of listening practices—reciprocity, humility, and

caution—that "aim to allow more listening by reducing the listener's aggression, that is, the speed and vehemence with which the listener's interpretive categories close in on the speaker's statements. Each strategy, then, involves some sacrifice of the listener's comfortable ground" (p. 13). Using the reciprocity stance, the listener privileges the speaker's perspective with the belief that the speaker is the most knowledgeable about her own experiences and ideas. The humility stance takes as an assumption that the listener has an incomplete understanding of the speaker's ideas. As Parker explains: "There is more that I must learn, and what appears to be a mistake on the part of the speaker would probably make more sense if I had a better grasp of the details, the emotions, the situation, and the speaker's history and social perspective" (pp. 13–14). Finally, the caution stance reminds the listener to move slowly before responding, making sure to avoid "denying or dismissing the validity of the speaker's point of view or manner of talking" (p. 14). Parker emphasizes that the purpose of these strategies is not to avoid disagreement but instead to foster a more equitable, productive exchange.

While we in the fields of early childhood and middle grades education may not always have opportunities for face-to-face conversations, these general strategies for listening seem applicable to the ways we might engage with each other both in person and through our writing and reading. Thus, we might begin our dialogues by identifying the common problem we seek to understand better and address—our shared commitments to and concerns about particular groups of young people, groups who have been historically misrepresented or neglected. We can then use Parker's strategies to make sure we are listening—or reading—with reciprocity, humility, and caution so that we might actually generate new ways of thinking and acting that take advantage of the diverse perspectives we have access to in our fields.

In order to keep our fields alive and keep them from being stagnant, a healthy skepticism about the foundational concepts, perspectives, and practices in the fields needs to be appreciated and actively sought. Topics such as the conceptualization of development, the role of culture, conditions/methods/benefits and problems of essentializing young children and young adolescents might provide good starts for conversations between scholars with different perspectives.

By recognizing different assumptions as well as shared commitments and goals among scholars with differing orientations, we may take a first step toward creating "agonistic pluralism" (Moss, 2007, p. 234)—a form of unity that does not deny different perspectives, but rather helps people with different orientations achieve "some form of engagement without requiring domination by one camp or a phony consensus" (p. 235). In doing so, we can build the alliance that Beane (2005) advocates and make the borderland path that we walk not only more inclusive and welcoming but also more likely to lead to a place of positive change in the education of young people.

REFERENCES

Beane, J. A. (2005). Foreword. In E. R. Brown & K. J. Saltman (eds), *The critical middle school reader* (pp. xi–xv). New York: Routledge.

Bloom, B. (1964). *Stability and change in human characteristics*. New York: Wiley.

Braun, S. J., & Edwards, E. P. (1972). *History and theory of early childhood education*. Belmont, CA: Wadsworth.

Bredekamp, S. (1987). *Developmentally appropriate practice in early childhood programs serving children from birth through age 8*. Washington, DC: National Association for the Education of Young Children.

Bredekamp, S., & Copple, S. (1997). *Developmentally appropriate practice in early childhood programs* (revised ed.). Washington, DC: National Association for the Education of Young Children.

Bruner, J. (1996). *The culture of education*. Cambridge, MA: Harvard University Press.

Bruner, J., & Haste, H. (1987). (Eds), *Making sense: The child's construction of the world*. New York: Routledge.

Cannella, G. S. (1997). *Deconstructing early childhood education: Social justice and revolution*. New York: Peter Lang.

———. (2005). Reconceptualizing the field of early childhood education: If "western" child development is a problem, then what do we do? In N. Yelland (ed.), *Critical issues in early childhood education* (pp. 17–39). Columbus, OH: Open University Press.

Cole, M. (1996). *Cultural psychology: A once and future discipline*. Cambridge, MA: Harvard University Press.

Conklin, H. G. (2008). Promise and problems in two divergent pathways: Preparing social studies teachers for the middle school level. *Theory and Research in Social Education, 36*(1), 591–620.

Copple, S., & Bredekamp, S. (2009). *Developmentally appropriate practice in early childhood programs serving children from birth through age 8* (3rd ed.). Washington, DC: National Association for the Education of Young Children.

Corsaro, W. A. (1997). *The sociology of childhood*. Thousand Oaks, CA: Pine Forge.

Damon, W. (1998). Preface to the handbook of child psychology, fifth edition. In W. Damon (series ed.) & R. M. Lerner (vol. ed.), *Handbook of child psychology: Vol. 1. Theoretical models of human development* (5th ed., pp. xi–xvii). New York: John Wiley & Sons.

Dillon, S. (December 16, 2008). Obama pledge stirs hoe in early education. *New York Times*. Retrieved August 25, 2009, from http://www.nytimes.com/2008/12/17/us/politics/17early.html.

Donaldson, M. (1978). *Children's minds*. New York: W. W. Norton.

George, P. S., & Alexander, W. M. (1993). *The exemplary middle school* (2nd ed.). New York: Harcourt Brace.

Hunt, M. (1961). *Intelligence and experience*. New York: Ronald.

Jackson, A., & Davis, G. (2000). *Turning points 2000: Educating adolescents in the 21st century*. New York: Teachers College Press.

James, A., Jenks, C., & Prout, A. (1998). *Theorizing childhood*. New York: Teachers College.

Kitayama, S., & Cohen, D. (Eds). (2007). *Handbook of cultural psychology*. New York: The Guildford Press.

Kliebard, H. M. (1995). *The struggle for the American curriculum: 1893–1958* (2nd ed.). New York: Routledge.

Lee, K. (2001). *Raising the independent self: Folk psychology and folk pedagogy in American early schooling*. Unpublished doctoral dissertation, University of Illinois, Urbana-Champaign.

Lee, K., & Walsh, D. J. (2001). Extending developmentalism: A cultural psychology and early childhood education. *International Journal of Early Childhood Education, 7,* 71–91.

Lerner, R. M. (2006). Developmental science, developmental systems, and contemporary theories of human development. In Damon & Lerner (eds), *Handbook of child psychology: Vol. 1.* (6th ed., pp. 1–16).

Lesko, N. (2005). Denaturalizing adolescence: The politics of contemporary representations. In Brown & Saltman (eds), *The critical middle school reader* (pp. 87–102).

Lexmond, A. J. (2003). When puberty defines middle school students: Challenging secondary education majors' perceptions of middle school students, schools, and teaching. In P. G. Andrews & V. A. Anfara, Jr. (eds), *Leaders for a movement: Professional preparation and development of middle school teachers and administrators* (pp. 27–52). Greenwich, CT: Information Age.

Lipsitz, J. (1977). *Growing up forgotten: A review of research and programs concerning early adolescence*. Lexington, MA: D. C. Heath and Company.

Mallory, B. L., & New, R. S. (Eds). (1994). *Diversity & developmentally appropriate practices: Challenges for early childhood education*. New York : Teachers College Press.

Moss, P. (2007). Meetings across the paradigmatic divide. *Educational Philosophy and Theory, 39,* 229–245.

Nourot, P. M. (2005). Historical perspectives on early childhood education. In J. L. Roopnarine & J. E. Johnson (eds), *Approaches to early childhood education* (4th ed., pp. 3–43). Upper Saddle River, NJ: Prentice Hall.

Parker, W. C. (Forthcoming). Listening to strangers: Classroom discussion and political education. In "Listening in Context," special issue of *Teachers College Record*, S. Haroutunian-Gordon and L. Waks (guest eds).

Rogoff, B. (2003). *The cultural nature of human development*. New York: Oxford University.

Ryan, S. K., & Grieshaber, S. (2005). Shifting from developmental to postmodern practices in early childhood teacher education. *Journal of Teacher Education* 56, 34–45.

Saltman, K. J. (2005). The social construction of adolescence. In Brown & Saltman (eds), *The critical middle school reader* (pp. 15–20).

Shepard, L. A. (1999). The influence of standardized tests on the early childhood curriculum, teachers, and children. In B. Spodek & O. N. Saracho (eds), *Issues in early childhood curriculum* (pp. 166–189). Troy, NY: Educator's International Press, Inc.

Shweder, R. A., Goodnow, J., Hatano, G., LeVine, R. A., Markus, H., & Miller, P. (1998). The cultural psychology of development: One mind, many mentalities. In Damon & Lerner (eds), *Handbook of child psychology: Vol. 1* (5th ed., pp. 865–937).

Stipek, D. (2006). No Child Left Behind comes to preschool. *The Elementary School Journal*, *106*, 455–465.

Walsh, D. J. (1991). Extending the discourse on developmental appropriateness: A developmental perspective. *Early Education and Development*, *2*, 109–119.

Wolfe, J. (2000). *Learning from the past: Historical voices in early childhood education*. Mayerthorpe, Alberta: Piney Branch Press.

Yeager, E., & Wilson, E. (1997). Teaching historical thinking in the social studies methods course: A case study. *Social Studies, 88*(3), 121–127.

CONTRIBUTORS

Sarah Bridges-Rhoads is a doctoral student in elementary education in the Department of Elementary and Social Studies Education at The University of Georgia. Sarah taught fifth grade in Athens, Georgia. Her interests include critical and post-structural theories relating to elementary, middle, and teacher education. Sarah's research has been published or is forthcoming in *Language Arts, Connections: NSRF Journal,* and in an edited book published by the International Reading Association entitled *Teachers as Readers: Perspectives on the Importance of Reading in Teachers' Classroom Lives.* In addition, she has presented her work at conferences such as the National Reading Conference and Curriculum & Pedagogy Conference.

Enora R. Brown is associate professor in the Department of Educational Policy Studies and Research in the School of Education at DePaul University in Chicago. She is the author of publications on the subjects of urban school policy and practice, youth and adult identity formation, human development, and middle school. She is coeditor with Kenneth Saltman of *The Critical Middle School Reader.*

Hilary G. Conklin is assistant professor of social studies education at DePaul University. A former middle grades social studies teacher, Hilary currently conducts research in the preparation of middle school social studies teachers and the pedagogy of teacher education. Her publications include articles in *American Educational Research Journal, Harvard Educational Review,* and *Theory and Research in Social Education,* and coauthored chapters with Ken Zeichner in *Studying teacher education: The report of the AERA Panel on Research and Teacher Education* and the *Handbook of research on teacher education: Enduring issues in changing contexts* (3rd edition).

M. Elizabeth Graue is professor of curriculum and instruction in the School of Education at the University of Wisconsin-Madison. She directs the Wisconsin Doctoral Research Program and is associate director of faculty, staff & graduate development at the Wisconsin Center for Education Research. A former kindergarten teacher, her areas of interest include school readiness, class size reduction, preparing teachers for inclusive home-school relations, and qualitative research methods. Her work has been published in numerous journals, including *American Educational Research Journal, Early Education & Development, Educational Researcher, Harvard Educational Review, Teachers College Record, and Teaching & Teacher Education.* She is the author of a book titled *Ready for what? Constructing meanings of readiness for kindergarten,* and has coauthored (with Daniel J. Walsh) *Studying children in context: Theories, methods & ethics.*

Lisa Harrison is a doctoral student in middle grades education in the Department of Elementary and Social Studies Education at The University of Georgia. Lisa taught middle school mathematics in New York City. Her research interests include African American adolescents' racial identity construction, incorporating social justice issues in middle grades mathematics curriculum, and African American adolescents' mathematical achievement. Lisa has presented her work at the annual meetings of the American Educational Research Association and National Middle School Association.

Hilary E. Hughes is a doctoral student in middle grades education in the Department of Elementary and Social Studies Education at The University of Georgia. A former seventh and eighth grade language arts teacher in Colorado, Hilary's research interests include social justice topics in middle grades and teacher education, literacy, and adolescent girls' identities. Her work has been published or is forthcoming in *English Journal, Middle School Journal,* and *Colorado Language Arts Society's Journal, Statement.* In addition, Hilary has presented at the annual meetings of the American Educational Research Association, National Reading Conference, and National Middle School Association.

Kyunghwa Lee is associate professor in early childhood education at The University of Georgia. A former kindergarten teacher from Korea, Kyunghwa explores cultural psychology as an alternative developmental framework of the twentieth century's grand developmental theories. She utilizes ethnography and cross-national research to examine various sociohistorical constraints that both support and hinder teaching and learning. Her work has been published in various journals, such as *Early Childhood Education Journal, Early Childhood*

Research & Practice, Educational Technology Research and Development, International Journal of Early Childhood Education, Journal of Early Childhood Teacher Education, Journal of Early Childhood Research, and *Journal of Research in Childhood Education.* She has guest-edited (with Stacey Neuharth-Pritchett) a special issue, Attention Deficit/Hyperactivity Disorder across cultures: Development and disability in contexts, of *Early Child Development and Care.*

Su Kyoung Park is a lecturer at Kyungwon University in South Korea University in South Korea. She earned her PhD in early childhood education from The University of Georgia. A former preschool teacher in Korea, Su Kyoung is interested in understanding early childhood teachers' lives. Her dissertation study explored the lives of induction teachers by paying attention to how these teachers negotiate their experiences within a specific teaching context and how they construct and reconstruct meanings of their teaching career. She has presented her study at the annual meetings of the American Education Research Association and the Georgia Science Teachers Association.

Amy Noelle Parks is assistant professor in early childhood education at The University of Georgia. A former elementary teacher, she earned her PhD from Michigan State University in curriculum, teaching, and policy. Her research is concerned with equity, mathematics education in early childhood, and the social contexts of schooling. Her work has been published or is forthcoming in *Teaching and Teacher Education, Teachers College Record, Journal of Curriculum Studies, Phi Delta Kappan,* and *For the Learning of Mathematics.*

Mark D. Vagle is assistant professor of middle grades education in the Department of Elementary and Social Studies Education at The University of Georgia. Before working in higher education, Mark taught in elementary and middle schools and was a middle school assistant principal. Mark received his PhD in curriculum and instruction from The University of Minnesota-Twin Cities. His current research and teaching can be broadly described as a continual pursuit of the dynamic interaction (tension, struggle) between relational and technical dimensions of pedagogy, especially as such dimensions are lived in context and over time. To this end, Mark draws on critical perspectives and uses phenomenology in his research. Mark's work has been published or is in press in the following: *International*

Journal of Qualitative Studies in Education, Teachers and Teaching: Theory and Practice, Handbook of Research in Middle Level Education (volume 6), Research in Middle Level Education Online, National Reading Conference Yearbook, Education and Culture, and *Pedagogies: An International Journal.*

INDEX

58
p252, 253
254
206 > BOOK
p230